Zooicide
Seeing Cruelty, Demanding Abolition

Sue Coe

With an essay by
Stephen F. Eisenman

Zooicide: Seeing Cruelty, Demanding Abolition

© 2018 Sue Coe, with Courtesy of Galerie St Etienne, NY.
Essay © 2018 Stephen F. Eisenman
This edition © 2018 AK Press (Chico, Edinburgh)

ISBN: 9781849352864
E-ISBN: 9781849352871
Library of Congress Control Number: 2018932258

AK Press	AK Press
370 Ryan Ave. #100	33 Tower St.
Chico, CA 95973	Edinburgh EH6 7BN
United States	Scotland
www.akpress.org	www.akuk.com
akpress@akpress.org	ak@akedin.demon.co.uk

The above addresses would be delighted to provide you with the latest AK Press distribution catalog, which features books, pamphlets, zines, and stylish apparel published and/or distributed by AK Press. Alternatively, visit our websites for the complete catalog, latest news, and secure ordering
Cover design by Suzanne Shaffer
Printed in the USA

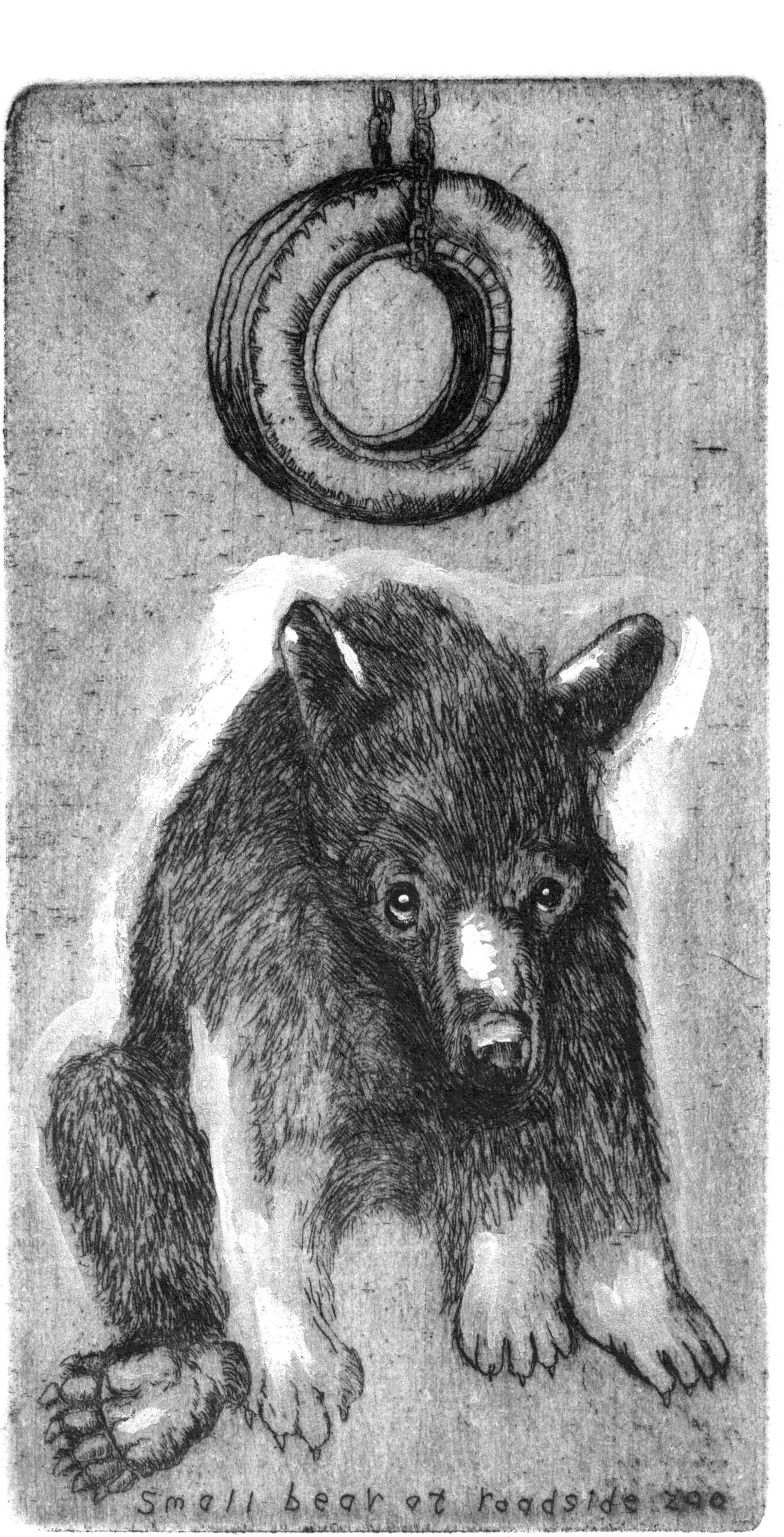
Small bear at roadside zoo

THE CAPITALIST ZOO

Looking Backward: 2050–1850

Strolling the London Zoological Refuge today, it's hard to believe the animals here were once imprisoned in cages or glass enclosures, treated as specimens or mere things, and prevented from living full and productive lives. But that was the system that existed for more than two hundred years after the London Zoo's opening in 1828. Now politics has changed, capitalist zoos have disappeared, and few people would patronize an institution that denied animals their rights.

In the United States, where zoos appeared later than England (the first was established in 1874 in Philadelphia), the treatment of animals was no better. From the beginning, animals were captured in the wild by buccaneers who cared little about how many died in the process. (It was routine to kill mothers to harvest babies.) Zoo animals were also obtained from unscrupulous brokers and bought from private menageries, many of which traveled the country alongside circuses. Even the largest and best zoos, such as the ones in New York (established 1899), Saint Louis (1914), and San Diego (1916), kidnapped animals from the wild, mistreated them, killed them, or drove them to zooicide. And change came slowly. As recently as 2018, the San Diego Zoo maintained a herd of four elephants on just two acres, and thirteen more on a parcel of just over six. (In the wild, elephants are nomadic, with a range of between five and twenty-five square miles.) In colder climates, zoo elephants were confined for much of the year in enclosures barely larger than themselves. The same was true of lions, tigers, gorillas, and many other of the most popular zoo attractions.

Would any institution today even consider doing what the Copenhagen Zoo did in 2014? Or what the St. Louis and Cincinnati zoos did in 2017? The first killed a healthy, eighteen-month-old giraffe named Marius, in addition to four lions to prevent them inbreeding. And in a further expression of conceit, they fed Marius to some lions while zoo visitors watched. The second, St. Louis, bred its elephants despite the tuberculosis and herpes that were endemic in the herd; and the third, Cincinnati, notoriously shot and killed a gorilla named Harambe who picked up a child who had fallen into his enclosure. Incidents such as these in which keepers killed animals in their charge—also called zooicide—were shockingly numerous.

To better understand the historic cruelty of zoos during the more than two centuries leading up to what was called "The Change," it's necessary to see them in a wider context. Large-scale animal agriculture in England in the mid-eighteenth century, turned animals from sentient beings into commodities—and from then on, their exploitation was accelerated. By 1850, millions of acres for grazing were opened up in the American West, Australia, and Argentina, with fattened animals transported from the American Great Plains, the antipodean woodlands, and the Pampas to stockyards and mechanized slaughterhouses. Cows and sheep became industrial raw materials. By the 1960s, concentrated animal feeding operations ("factory farms") were the norm for meat production in the U.S., and many other rich countries across the globe. This was the age of McDonald's, when children recognized Big Macs and McNuggets more easily than they did living cows and chickens. This was the era of "cattle futures," when investors bet on whether the value of "feeder cattle" (eight-hundred pounds each, well-muscled) would rise or fall in a given time frame. No actual cattle had to change hands—the animal was merely an abstraction, an instrument for the creation of profit. Of course, all living cattle were, in the end, slaughtered.

If that wasn't bad enough, governments everywhere in the twentieth and early-twenty-first centuries, organized senseless pogroms—in the name of sanitation and ecological control—against rats, rabbits, badgers, possums, coyotes, kangaroos, deer, and other so-called "pests" or "invasive" species. In addition, innumerable monkeys, mice, and fish were tormented and killed in biomedical research laboratories in the U.S., Europe, China, and elsewhere, despite the proven failures of animal testing for determining the usefulness of medicines. The profits of drug companies and university research labs were simply too high for them to consider changing basic research protocols. In the name of fashion, men and women during these decades frequently wore the flesh and fur of animals. They also used them to cover the seats of their gasoline-powered cars. Millions more people participated in blood sports such as hunting, horse racing, dogfighting, rodeos, bullfighting, and cockfighting.

During wartime, zoo animals were victims of prejudice, superstition, and sheer malice. In World War II, civilian and military leaders in London, Berlin, and Tokyo killed many zoo animals out of the unfounded fear they would escape during aerial bombing and terrorize the population. At the Ueno Zoo in Tokyo, the elephants were killed in order to demonstrate to the Japanese population that wartime demanded sacrifices from everyone, zoo animals included. They were starved to death. In 1992, the last animal at the Sarajevo Zoo—a female Black Bear—died of starvation after eating the carcass of her mate. The fate of zoo animals in the twentieth century—always bad—became terrible in wartime.

Even artists, themselves often mistreated by the wider society, sometimes visited terrible cruelties upon animals in the name of originality. It was a perverse enterprise at best: What, at that time, was more conventional than animal abuse? And what, in the end, distinguished a dead shark in a vitrine by Damien Hirst (*The Physical Impossibility of Death*, 1991), from a living one in a tank at the Monterey Bay Aquarium? (Sharks in captivity die very quickly.) How does *Dogs Which Cannot Touch Each Other* (2003), a video of pit bulls on treadmills straining at each other, by Peng Yu and Sun Yuan, differ from a staged dogfight? In both cases animals are taunted into a fury for the benefit of a human audience. Only a few artists at the time dedicated their work to the protection of animals. The most prominent of them was the English-born, American artist Sue Coe. She saw and depicted almost every horror visited upon animals in the conviction that cruelty must be witnessed and represented.

The early-twenty-first century marked the very nadir of animal rights. The rate of animal extinctions rose alarmingly, perhaps as much as 10,000 times the background rate. Between about 1970 and 2020, the number of fish in the oceans declined by 50 percent; large mammals in the wild by 66 percent; global insects by 75 percent. But nature has a way of exacting revenge, and sure enough, the release into the atmosphere of carbon dioxide and methane from industry, transportation, and animal agriculture warmed the climate and caused more intense rainstorms and sea level rise, inundating major cities. Animal waste from factory farms entered into lakes, rivers, bays, and even oceans, poisoning them. The decline of insects (the result of reckless pesticide and herbicide use), and consequent loss of pollinators, damaged fruit, nut, and vegetable harvests. The planet itself was becoming unlivable for humans, in significant part because of the ongoing war against animals.

But by the end of the 2020s, after the final overthrow of the Trump regime in the U.S. and a series of hurricanes, floods, and fires of unprecedented scale, growing numbers of people in most countries began to accept both the moral worth of animals and the environmental peril caused by killing and consuming them. The prospect of immanent destruction seemed at last to steer the human, ethical compass. Thus, the change began. Vivisectors confronted ever-increasing social ostracism. Hunting declined everywhere, and popular protests curtailed extermination campaigns against supposed pests and invasive species. Veganism grew from a fringe movement to an unstoppable social and political force. And the movement for animal protection in the twenty-first century was propelled as well by that peculiar institution found in cities all across the world in addition to London: that is, zoos.

The irony is that zoos were important in securing rights for animals, not because they so successfully carried out their stated education and conservation missions, but because they failed so miserably at them. The growing, popular recognition that animals are conscious beings with the same moral rights to life and liberty as people, meant that imprisoning them, infantilizing them, treating them as commodities, driving them to madness or suicide, and sometimes even killing them in zoos was now insupportable. Beginning in the 2020s, organized boycotts disrupted the operation of hundreds of zoos around the world. Increased environmental regulation and controls of capital—required to stop the release of greenhouse gases and save human life on the planet—prevented corporations from marketing themselves through zoo sponsorships. And the withdrawal of corporate funding, combined with the boycott movement, led to greatly reduced zoo attendance. The only possible future for zoos, it quickly became clear, was as clinics, way stations, research facilities, and refuges. If not that, then they must disappear entirely. And little by little, zoos in the traditional sense were no more, facilitators of the widespread movement for animal liberation and near universal veganism.

What began in the early-nineteenth century as popular entertainments with a veneer of science—animals in cages placed in public parks—as in London, became full-fledged research, protection, and education facilities by the middle of the twenty-first. The barter and forced breeding of animals was ended. The zooicide of supposed excess animals ceased. Exotic animals were placed in zoo-refuges only if their lives depended on it, and then for as short a time as possible before they were returned to the wild or to a larger wildlife sanctuary in open country. (For a short while, in the early-twenty-first century, traditional zoos began calling themselves "sanctuaries" or "conservation centers" without making any other changes, but that ruse was soon exposed.) A visit to the zoo no longer meant looking down on animals as if they were children, servants, prisoners, trophies, specimens, slaves, or things; it meant seeing them as individual beings with moral worth and supporting the effort to improve their conditions of life. Zoos today, in London, Chicago, New York, Beijing, and elsewhere, contain orphaned, exotic animals awaiting placement with appropriate animal families in protected habitats; injured urban and semi-urban animals such as pigeons, squirrels, rabbits, and coyotes; and the offspring of formerly domesticated animals, including chickens, cows, sheep, and pigs. Zoo attendance is higher than it's ever been, as visitors are enabled to discretely observe and admire the treatment and recovery of injured or orphaned animals, walk freely among happy, domesticated animal families, and watch holographic and virtual-reality projections of once endangered, but now thriving, exotic animals in their native habitats.

But the passage from one form of zoo to the other was only possible because of a radical shift in people's attitudes toward animals, each other, and nature. It required them to resist the acquisitive posture of political and corporate leaders who measured the world and its creatures according to profit and loss, and to embrace instead the attitude of poets, naturalists, environmentalists, and animal rights campaigners like William Blake, Henry David Thoreau, Rachel Carson, Tom Regan, and Gary Francione:

> A Robin Red breast in a Cage
> Puts all Heaven in a Rage
> A Dove house fill'd with Doves & Pigeons
> Shudders Hell thr' all its regions
> A dog starv'd at his Masters Gate
> Predicts the ruin of the State
> (William Blake, "Auguries of Innocence," 1806)

No humane being, past the thoughtless age of boyhood, will wantonly murder any creature which holds its life by the same tenure that he does. The hare in its extremity cries like a child. I warn you, mothers, that my sympathies do not always make the usual phil-*anthropic* distinctions.
(Henry David Thoreau, *Walden, or Life in the Woods*, 1854)

Until we have the courage to recognize cruelty for what it is—whether its victim is human or animal—we cannot expect things to be much better in this world. We cannot have peace among men whose hearts delight in killing any living creature.
(Rachel Carson, "Letter to Fon Boardman," c. 1952)

Being kind to animals is not enough. Avoiding cruelty is not enough. Housing animals in more comfortable, larger cages is not enough. Whether we exploit animals to eat, to wear, to entertain us, or to learn, the truth of animal rights requires empty cages, not larger cages.
(Tom Regan, *Empty Cages*, 2005)

The idea that it's educational to observe animals in cages tells us all we need to know about how deeply embedded our speciesism really is. We purport to learn about who they are as we gawk at them in their confinement. We take our children to zoos to ensure that the next generation continues to think that nonhuman animals are merely things that exist for us, and whose natural state is to be reduced to being our prisoners. There are no good zoos. There is no compassionate exploitation.
(Gary Francione, letter to the author, April 2018)

These pioneers and many others, understood humans and animals, as well as culture and nature to be co-dependent and of equal value. Humans are themselves animals, they reasoned, and an injury to the one is an injury to the other. Animal rights are therefore also human rights. In addition, human consumption and production must be mutually supporting if the environment for humans and animals is to be conserved. The waste from consumption must become a resource for production. And most of all, industry and manufactures must be deployed for the sake of need and delight, not for the profit of the wealthiest and most powerful. When these radical perspectives began to gain traction, the entire exploitative model that underlays capitalist zoos began to crumble, and a new chapter was begun in the long story of humans and animals.

The History of Human and Animal Relations, Briefly Told

The history of human and animal relations may be divided into three ages: *Communal*, *Tributary*, and *Capitalist*. During the first, communal period, from roughly 100,000 to 10,000 BCE, humans lived together as foragers in relative peace, without much social hierarchy. Wild animals roamed or resided alongside them and were considered by humans to be conscious beings with rights and obligations similar to theirs. Zoos were unknown during this era, though by the end of it, and for reasons unknown, some animals were forcibly herded into pens or camps, initiating both domestication and widespread human aggression toward animals.

To describe the communal period and forager society as representing an Edenic or Golden Age in human-animal relations would be a romantic exaggeration however. For one thing, humans hunted animals, and in a few cases appear to have eradicated whole species. The Paleo-Indian Clovis people, for example, hastened the extinction of the elephant-like Gomphotheres more than 13,000 years ago. Other large animals too went extinct during the late Pleistocene and early Holocene eras, probably from human predation, including the Woolly Mammoth, Irish Elk, and Giant Tapir. But the relatively small scale of human society at that time, the absence of sophisticated hunting tools, widespread belief in the moral being of animals, and notions of a required reciprocity between them, limited exploitation. If communal society had remained intact, the world would not be dominated—and perhaps doomed—as it is today, by a single, two-legged, particularly rapacious species.

In the second, tributary age, extending from about 10,000 BCE to 1800 CE, large-scale civilizations arose, and social hierarchies and inequality grew. Menageries (collections of exotic animals) were established by wealthy individuals and families and deployed as overt expressions of power and status. Though humans dominated animals in the course of these millennia, and the ideology of "speciesism" (irrational belief in human superiority) was born, animals were still generally considered beings who experienced pleasure and pain and had distinct interests and preferences. They were also believed to embody or at least symbolize human virtues and vices, and their lives and deaths had more than just commercial significance. However, the rapid growth of huge markets for live and dead animals at the end of this period, for example at Smithfield in London and Poissy in Paris, marked the start of a new, more oppressive order.

Tributary society in general inherited many communal beliefs, including the idea that humans had a moral responsibility to animals, as well as that animals themselves had moral obligations. (In the late middle ages and early modern period, animals were sometimes charged with crimes.) But it also established a new, more troubled relationship with animals, both wild and domesticated. Animals were expropriated from nature through widespread hunting. They were also herded, concentrated in farms, and sold at market. Human domination of animals was overt and became codified in religion and civil law. Anthropocentrism—the idea that humans are the core of creation—went largely unquestioned.

Finally today, during the third, capitalist period, and especially under the aegis of so-called "neo-liberalism" (the current era) when a handful of global corporations and immensely wealthy individuals manage nearly the entire global economy and political order, civilization seems to be at war with nature itself, including non-human animals. Species after species has been either greatly reduced in number or threatened with extinction. Wild animal habitats have been replaced by grazing land, housing, or parking lots, and global warming has progressed faster than most species can adapt. Animals' only effective weapons of defense in this epoch, used without deliberation and with haphazard effect, have been the zoonotic diseases that kill millions. At the beginning of the capitalist period, between about 1780 and 1850, zoos in the modern sense arose as expressions of the general, Enlightenment impulse to rationalize, colonize, and expropriate the natural world and justify animal agriculture. Animals were thus robbed of their souls and their social being; they symbolized nothing but their biological selves—their genus and species—and were exhibited as entertainments or trophies in zoos all over the world.

During the most recent phase of the capitalist epoch, in the later twentieth and early-twenty-first centuries, many zoos changed their stated missions to focus on educating the public about the forces driving animals to extinction. In fact, however, zoos in this age upheld and even extended the widespread *thingification* of non-human animals—their treatment as objects without any history or culture of their own, and as having no worth other than as instruments of production or exchange. By enclosing animals in cages and considering them alternately as precious specimens and entertainments (especially for children) instead of as individual beings with moral significance, and by buying, selling, trading, breeding, and even killing them, zoos promoted the very anthropocentrism many of them claimed to be attacking.

The Birth of the Menagerie

Before the appearance of settled, agricultural civilization about 10,000 years ago, there were no zoos or menageries.

In communal society, among foraging people (also known as hunter-gatherers), the division between human and non-human beings was fluid, and ownership of the latter made no more sense than possession of the former. Foragers were generally at peace with each other and lived in relative harmony with non-human nature. They understood life to be a free-flowing force that spread across the world, pre-existing any individual living thing and able to assume any form. After death, a human might change into an animal or an animal into a human, preserving the animating life force. Thus, humans had respect for animals, even when hunting them, and felt a sense of kinship with them—after all, any animal might once have been a family member! Of course, the family is sometimes a cradle of violence, and some species suffered terrible losses, even to the point of extinction, at the hands of Paleolithic and Neolithic hunter-gatherers.

Based upon both ethnographic and ethno-historical evidence, foragers are less violent toward each other than are agriculturalists, and less likely to use or abuse animals. Kindness to animals was not simply a matter of custom, however. It was practical too, encouraging some animals, so hunters believed, to sacrifice themselves in order to help sustain foragers and their clans. Far from being commodities therefore—mere things to be bought and sold—animals in the communal period possessed moral autonomy. It would be unimaginable to confine them to cages.

Like foraging, pastoralism is a communal, economic system of great antiquity. Pastoralists subsist by tending herbivorous animals, shepherding them between grazing areas, and selling or trading individual members of the herd to other pastoralists or sedentary people for money or necessities. (Unlike foraging, which has become very rare, pastoralism is still practiced by millions of people, mostly in central Asia and the Sahel region of west Africa.) Though pastoralists enslave, sell, and slaughter individual animals, they do not deny their moral worth or treat them as mere things; the idea of killing or selling off large numbers of them to amass profit, subsequently invested to generate more profit, is anathema. In addition, herders may flaunt possession of particularly unusual or prized animals as expressions of status, as Evans Pritchard long ago described, but would never violate their species nature by confining them in cages for purposes of exhibition or entertainment. To do so would satisfy neither moral nor economic logic.

Nevertheless, menageries were born, in a sense, with the emergence of pastoralism. The first grazing animals collected into herds—ancestors of modern goats and sheep—were obviously untamed, and the results of domestication could not have been known at the start. Therefore, menageries of a rudimentary kind must have existed at the beginning of the Holocene, 10,000–12,000 years ago. In addition, human families or clans may have identified with particular animals and maintained small collections of them as displays of rank or status. Animals could symbolize natural forces, powerful gods, or historical change. Animals may also have been gathered together to better understand the distinction between human and animal lifeways—not so much a scientific enterprise as an informal, comparative one. All humans, regardless of the economy and society in which they live, are curious about animals—they look closely and think long and hard about them. In any event, these small menageries were the nuclei of the larger ones formed by the sedentary peoples of the Middle East, Central and East Asia, Africa, and Europe, as well as the larger herds tended by agriculturalists in the tributary age.

Menageries in the Tributary Age

The first purposeful collections of exotic animals were the product of tributary culture—societies in which economic transactions were transparent and oppression was undisguised. In these societies, goods were made and consumed by the producers themselves, who were forced to pay tribute (theft of labor, rent, or goods-in-kind) to a few, powerful, secular, or ecclesiastic rulers. This basic system, with many local variations, lasted for roughly 7,000 years between the rise of Egyptian and Mesopotamian civilization and the take-off of capitalist society in the mid-eighteenth century. The purpose of menageries in this context was to honor and legitimate existing authority so as to forestall challenges either from other elites or subservient classes. Like warfare, imprisonment, and the building of temples or pyramids, menageries were expressions of sovereignty. To possess, control, hunt, and exhibit exotic animals—especially predators such as lions and tigers—was to both display power over territory and assert a monopoly of violence. Menageries arose during the tributary age as symbols of domination.

The precise form and function of menageries in early Egypt, before it was united under the pharaohs, is unclear. But skeletal remains found at Hierakanopolis (3100 BCE)—including those of baboons, leopards, and elephants—indicates a large, exotic, animal population. They also reveal that the creatures were beaten and abused, probably in order to gain their submission or ensure their docility. Later Egyptian menageries may have been less abusive, but the animals in these were frequently hunted to death. By the late dynastic period, and especially after conquest by Alexander in the late-fourth century BCE, animals were integrated into Dionysian festivals as symbols of *luxuria*—the lust for sex, money, or power. At the *Ptolemaieia*, established by Ptolemy in 278 BCE, there were seen:

> twenty-four elephant chariots, sixty teams of he-goats, twelve of saiga antelopes, seven of beisa antelopes, fifteen of leucoryse, eight teams of ostriches, seven of Père David deer, four of wild asses, and forty-four horse chariots. On all of these were mounted little boys wearing the tunics and wide-brimmed hats of charioteers, and beside them stood little girls equipped with small crescent shields and wand-lances, dressed in robes and decked with gold coins (Athenaeus, *Deipnosophistae* Book V).

The parade ended with bears, leopards, genets, caracals, and most remarkable of all (because of their size and rarity) a rhinoceros and a giraffe. The greatest menagerie of antiquity was established at about this time in Alexandria, site of the great library. It was part of the effort to create a broad inventory of knowledge about and control over the natural world. But its scholarly purpose was exceptional and its lifespan was limited. When the Romans conquered the city three hundred years later, its zoo specimens became a resource for the gruesome gladiatorial and animal combats in the Colosseum.

Animal deaths in tributary menageries must have been legion. Capture itself probably killed most creatures. Many others would quickly have died from long transit, poor diet, sickness, and abuse. And still more animals, especially lions, were killed while hunting (an elite entertainment), or from being hunted. It is known that lions were kept in cages and pits as early as Ur III (2100 BCE), but the best evidence of captive hunts comes much later, from Assyria. Scenes of staged royal lion hunts, for example, appear in alabaster reliefs in Room C of the North Palace at Nineveh, built by Ashurbanipal from about 650 BCE. There we see the king in his chariot, accompanied by bodyguards, soldiers, and beaters (who rouse the animals), stab and shoot with arrows the lions collected for the purpose. The most disturbing reliefs show dead or dying lions, [BELOW] including one of a lioness raising her head in a last roar as she drags her useless rear legs, paralyzed from the arrows lodged in her spine. The image displays great sensitivity and

Artist unknown, Panel from Room C of the North Palace at Nineveh, 645-635 BCE.

pathos. Ninevah at the time was one of the largest cities in the world, but its communal, pastoralist origins are apparent here: the lions are depicted as sentient beings possessing moral autonomy. They don't just experience pain; they are capable of moral suffering.

In ancient China, around 1000 BCE, Wen Wang, the first emperor in the Zhou dynasty, created a large, walled "Garden of Intelligence" or "Divine Park" that contained many exotic species, including antelope, deer, and numerous birds and fishes. It had a dedicated staff of keepers and veterinarians and was a source of both food and entertainment. Later, during the Han dynasty, many smaller menageries with caged animals were established, a tradition that continued down to the Yuan and Ming dynasties and beyond. These impacted later European menageries through the account of Marco Polo at the court of the first Yuan emperor, Kublai Khan:

> The Emperor hath numbers of leopards trained to the chase, and hath also a great many lynxes taught in like manner to catch game, and which afford excellent sport. He hath also several great Lions, bigger than those of Babylonia, beasts whose skins are colored in the most beautiful way, being striped all along the sides with black, red, and white. [Marco Polo obviously made no distinction between lions and tigers.] These are trained to catch boars and wild cattle, bears, wild asses, stags, and other great or fierce beasts. And 'tis a rare sight, I can tell you, to see those lions giving chase to such beasts as I have mentioned! When they are to be so employed the Lions are taken out in a covered cart, and every Lion has a little doggie with him. [The dogs were evidently companion animals to calm the tigers during transport.]

Greek and Roman collections of animals, like Persian and Arab ones somewhat later, were maintained on vast estates and in palaces. Few of these were seen by non-elite audiences, though individual animals or small groups might sometimes be paraded for public exhibition. These animal collections were not zoos in any modern sense, but transparent displays of elite wealth, status, and power, designed to ensure the loyalty or obedience of courtiers and visitors. These animals, like domesticated animals used for food or labor, were mere "inarticulate instruments," in the phrase of the Roman Marcus Terentius Varro, and due little moral consideration. In imperial Rome, hundreds of exotic animals were held in stockyards or *vivaria* before and after the public *venationes* during which they were made to fight gladiators or each other. These notorious spectacles, involving dozens or even hundreds of animals continued, though at a smaller scale, at Constantinople.

After the fall of Rome and during the European Middle Ages, menageries appeared in many court and ecclesiastic settings. In the eighth century, the emperor Charlemagne maintained several collections of exotic animals at his estates. So did William the Conqueror in England during the late-eleventh century. Henry III established a menagerie at the Tower of London in 1235 that included lions, an elephant (it survived about a year), and even a polar bear said to have swum and hunted for fish in the Thames. The artistic corollary of the medieval menagerie was the bestiary, quasi-encyclopedic compendia of animal effigies with fables, epigrams, and biblical stories intended to affirm scriptural truth and instruct humans about virtues and vices. In the Ashmole Bestiary (c. 1250), for example, a dragon is shown in one panel suffocating an elephant, just as Satan, the text informs us, waits to strangle humans with their own sins.

During the Renaissance, a period of self-conscious revival of the art, architecture, law, and customs of the Greco-Roman past, princes, lords, and nobles assembled menageries and enclosed hunting grounds, just as had the kings and emperors of antiquity. Many works of art in the Early Modern period, from about 1400 until the demise of tributary culture in the late 1700s, represent or symbolize kingship, hierarchy, and moral law (the contest of virtues and vices), through the depiction of animals. These virtual menageries, the artistic mirror of real ones, are summations of roughly ten millennia old ideologies of speciesism, anthropocentrism, and human exceptionalism—the last is the idea that humans exist outside of nature and possess a God-given authority to command and consume anything in the world they want, without consequences. These perspectives arose in the course of the tributary period; they gradually replaced, though not entirely, the perspective dominant in communal societies: that animals were beings who possessed moral worth.

Piero di Cosimo's painting, *A Hunting Scene* ([PAGE 8, TOP], c. 1494–1500), is an expression of early modern speciesism and a veritable menagerie. It depicts no less than thirty animals, not including several centaurs (half man and half horse) and satyrs (half man and half goat). In the foreground, a male lion seizes a brown bear who in turn is attacked by a black bear restrained by a human wearing an animal skin. Elsewhere in the picture, a white dog attacks a female lion, a bear chases a stag, a man subdues a bear by gripping its muzzle, and three other men carry off

Piero di Cosimo, *A Hunting Scene*, 1494–1500.

an ox they have killed. At the lower right, in extreme foreshortening, is a bloodied, naked human body, victim of the battle of man and beast. A small monkey in a tree at middle right, gazes upon the carnage, while forest fires blaze in the distance.

The meaning of the picture and its apparent companion, *The Forest Fire* (Oxford, Ashmolean Museum), has never been precisely determined, but its general idea—widespread in antiquity—is that human society developed from barbarism to civilization though the mastery of other animals and the conquest of fire. At first, humans were little more than beasts themselves, Lecretius writes: "And by the aid of their wonderful powers of hand and foot they would hunt the woodland tribes of beasts with volleys of stones and ponderous clubs, overpowering many, shunning but a few in their lairs; and when night overtook them, like so many bristly hogs they just cast their savage bodies naked upon the ground, rolling themselves in leaves and boughs" (*De Rerum Natura*). But soon, humans discovered fire, which permitted them to create domestic life, marriage and legitimate children, as well as establish natural hierarchies: "Kings began to found cities and to build citadels for their own protection and refuge." The painting clearly portrays human ruthlessness toward non-human animals, but at the same time shows humans themselves as a species of beast. They are naked or draped in animal fur, and succeed by sheer force of numbers, not by any manifest guile.

The Forest Fire [BELOW] is even more ambivalent in its triumphalism, though many mysteries about it too remain. It shows dozens of animals, this time however, they are fleeing a very destructive fire. The man at middle right carries a yoke over his shoulders as if he were an ox, while a pair of animals at middle left (a boar and a deer) have human heads, further blurring the distinction between human and animal. Whereas the Roman Vitruvious (*De Architettura*) saw fire as the basis of human society, speech, and even architecture, Piero appears less sure—fire drives all creatures, human as well as non-human, from their forest dwellings and enforces a community of man

Piero di Cosimo, *The Forest Fire*, 1505.

The Capitalist Zoo

and beast. For Piero, and presumably too for his small circle of patrons in Florence and Rome, animals possessed at least a modicum of moral being—they were not simply "inarticulate instruments."

In addition to moral being, animals in the tributary age possessed meaning—they were symbols, for example, of virtues and vices, and were represented as such in medieval bestiaries, as we have seen, fables, emblem books, and epics. At the start of Dante's *Divine Comedy*, the narrator is lost in a dark forest, which represents sin. There he is threatened first by a leopard (*Luxuria*/lust), then by a lion (*Superbia*/pride), and finally a wolf (*Avaritia*/avarice). These vices however, are not foreign to humans; they are basic to them. The animal resides, in a sense, within the human and the human within the beast. The same animal symbolism is represented in Titian's *Allegory* ([BELOW], London, National Gallery, c. 1565), where the dog however, replaces the leopard. (Changes in the roles of animals are common in the allegorical tradition.) A human head, corresponding to the "three ages of man"—young, middle aged, and old—surmounts each. Painted near the end of the artist's life, the painting may represent, as Simona Cohen has suggested, a self-accusation and confession. The old man at left is a self-portrait, his expression exaggerated to resemble the wolf beneath. The image is thus a penitential expression of the sin of avarice and the moral proximity of man and wolf.

But the recognition within tributary society of animals as both moral beings and symbols diminished over time as property relations changed. There arose in the early modern period a new attitude toward the use and ownership of animals of all kinds, domesticated as well as exotic. From the serfs of the Middle Ages sprang the farmers, minor land-holders, and small-scale traders of the earliest towns. From them in turn, developed the nascent capitalist class, sometimes called the bourgeoisie. They collected rent, engaged in manufacturing, trade, and the professions, and developed animal agriculture. These activities often operated in combination. For example, sheep's wool could be spun into thread, woven into fabric, sewn into garments, and sold locally or shipped abroad.

For the woolen industry to succeed, selected animals—enclosed within walls, fences, or natural barriers—were bred in order to improve the productivity of the stock and increase the value of their wool and meat. In ever larger numbers, cows and pigs too were bred and taken to large markets to be sold for money or to be butchered, enabling the purchase of still more animals, further refinement of breeds, new sales, and so on, with the goal always being profit and ever larger trade, not direct consumption. By the end of the tributary period, the feudal system of production—in which industry was monopolized by closed guilds, and livestock by small-scale farmers and butchers—no longer satisfied the growing wants of the new markets and industries. Guild-masters were pushed to one side by the manufacturing middle class, and the division of labor among the different corporate guilds vanished in the face of division of labor in each single workshop. In the New World, vast frontiers allowed for the establishment of ranches where pigs, cattle, sheep, oxen, and horses were exploited for their meat, hides, wool, and labor, the latter of which enabled still other industries to grow including sugar and cotton. The last two were abetted by slavery—through which humans were owned and abused just as animals were. Single herds of animals sometimes numbered in the tens of thousands; and the quantity of animals killed annually—for example cattle in Argentina in the 1820s—in the hundreds of thousands.

In such a context, the form and meaning of menageries must change, and they did. From expressions of noble and aristocratic power, menageries became laboratories for the acclimatization of new species that might be profitably exploited, as well as entertainments for the middling and lower classes of Europe. From manifest expressions of the Great Chain of Being (the idea that all creation exists in a ranked order from worms to God), they became visual tokens of new, rationalizing systems of taxonomic classification. (The model that triumphed was Linnaeus's binomial one, comprising genus

Titian, *Allegory of Prudence*, 1565-1570.

and species.) And from living embodiments of virtue and vice, animals became spectacles beyond good and evil. As such, the animals in zoos changed from *beings* (however vitiated) and *symbols* (however changeable) to mere *things* to be bought, sold, traded, bred, exhibited, and exploited. Most nineteenth and twentieth century zoos—let's call them Capitalist Zoos—were profit driven. But even when they were not, their function was to create profit in other sectors, for example, real estate, or else to provide policed leisure for workers and their families, within the domain of what is sometimes called *social reproduction*—the production and maintenance of the producers. (To this day, mothers and children are the largest population of zoo visitors.) Finally, the capitalist zoo also functioned to normalize industrial-scale, animal agriculture. If one set of animals (exotic) is fit only for watching, the other set (domesticated) must be fit only for eating.

The Capitalist Zoo

Menageries are called zoos when they are organized according to scientific principles. A number of such zoos—in Vienna, London, and Paris—appeared in the late-eighteenth and early-nineteenth centuries when nature was increasingly rationalized and systematized. These zoos were mainly organized by taxa—for example, there were separate enclosures for monkeys, cats, and birds. A little later, geography became the predominant organizing principle—animals were located in the zoo according to the continent or colony from which they came. More recently, zoos have organized their exhibitions, like many botanical gardens do, according to ecological zones: savannah, desert, mountain, rainforest, etc. Sometimes, all three are deployed. At the London Zoo, the Reptile and Amphibian house is organized by taxa, the Land of the Lions by geography (the former colony of Gujarat), and Rainforest Life by ecology.

But it's not so much their organizing principles as their capitalist basis that distinguishes modern zoos from pre-modern, or tributary menageries. Whereas menageries depended upon the expropriation (also called primitive accumulation) of wild animals, zoos required both the expropriation and exploitation of animals. It treated them as "instruments of production," to quote the Caribbean Aimé Césaire's famous *Discourse on Colonialism* (1955). Animals in zoos, like colonial subjects, are subject to *thingification*. They are exploited to create value.

The distinction between menageries and zoos may be simply expressed by means of three, general formulas:

Type A – The Pre-capitalist menagerie:
a-m-a
An animal (a) is taken from the wild and exhibited in a menagerie (m). When it dies, it is replaced by another (a). The purpose of acquiring an animal is to exhibit it.

Type B – The Industrial capitalist (early-nineteenth century) zoo:
z-a-z'
A zoo (z) is created by investors and stocked with animals (a), increasing its value (z'). The purpose of the zoo is to enhance investor-patron value.

Type C – The Neo-liberal (current) zoo:
z-a-z'-(h,d,r,m,s,b)-z"
A zoo (z) is created by investors or patrons and stocked with animals (a), increasing its value (z'). Natural-looking habitats, didactics, research, marketing, corporate sponsorships, and breeding are added (h,d,r,m,s,b), further enhancing its value ($z"$). The purpose of the zoo is to increase zoo revenue and create investor, patron, and corporate wealth.

Like menageries, zoos acquire animals by theft, purchase, trade, or breeding, and then keep them in cages or other enclosures. And while zoos developed from menageries, there are enormous differences between them, as the formula above suggests.

Type B zoos contained or anticipated some elements of Type C zoos. Consider the case of the Jardin des Plantes in Paris and "La Belle Africaine," a giraffe presented by Muhammad Ali Pasha, viceroy of Egypt, to King Charles X of France in 1827. When she arrived in Paris, dressed in a cape embroidered both with the French fleur-de-lis and the Ottoman crescent, she was the first live giraffe to appear in Europe in more than three hundred years, and the first ever in France. She quickly became a sensation. Not only did she advertise French efforts to control North Africa, she sparked giraffe fashions and giraffe profits for a decade and a half, until her death in 1845. By importing giraffes, lions, tigers, elephants, and other animals from its overseas colonies, France gained something of value, created value in other sectors, and publicly proclaimed its authority over the colonial territory from which the animals came. In 1850, Isidore Saint-Hilaire, zoologist son of the famous scientist who escorted the La Belle from Marseilles to Paris, wrote: "Through the…establishment of colonies in every corner of the globe," humanity takes possession of "species already in the power of other peoples." The expropriation of exotic animals, in other words, not only validated France's existing colonial authority, it potentially expanded it.

A decade later, Saint-Hilaire presided over the creation of the new Jardin Zoologique d'Acclimatation in the Bois de Boulogne, based upon the idea—endorsed by the Emperor Louis Napoleon himself—that some exotic animals (including antelopes, yaks, gazelles, and kangaroos) could be acclimated to a new climate and profitably exploited for their meat, fur, or hides. By the 1870s however, the Jardin Zoologique had proven unprofitable and the facility in the Bois was converted to a human zoo, an explicitly racist Jardin l'Acclimatation Anthropologique where Nubians, Inuit, Zulus, San, and other people were exhibited. Attendance quickly doubled. (That particular zoo closed in 1912.)

Another early example of zoological exploitation, profit, and colonialism is the famous Indian elephant Chunee in Regency London. Brought over from India in 1809, Chunee was exhibited first at Covent Garden and then at a small zoo, upstairs in the Exeter Exchange on The Strand. After years of relative placidity, he grew increasingly rebellious due to mistreatment, until finally in 1826—perhaps agitated by a sore tusk or by "musth" (a period of high testosterone levels)—he attacked and killed his keeper. Within days, a decision was made to execute him. After a failed attempt with poison, naval soldiers from Somerset House were summoned, and on March 1, Chunee was shot with 152 musket balls, an episode illustrated by Sue Coe [PAGES 26 AND 27]. His death by military firing squad, according to witnesses, was excruciating and protracted.

A hundred years later, George Orwell described a police officer in Moulmein Burma called upon to shoot a working elephant, also in musth, who had killed a villager. Orwell's narrator, modeled on the writer himself, opposed the British occupation and hated the thought of killing the animal. Moreover, the elephant's agitation appeared to be lessening by the time he'd been summoned. But surrounded by two thousand native villagers, the policeman decided he must act the "sahib" and shoot the elephant. Bullet after bullet put the animal "in great agony." Finally, after a half hour of suffering, the elephant died. He'd been executed for no other reason, the narrator states, than so the policeman could "avoid looking the fool." Both Chunee and Orwell's elephant were killed to demonstrate hegemony over unruly colonial subjects, both animal and human.

Even today, colonialism and hegemony, in addition to profit, are fundamental to zoos and the exhibition of animals. The London Zoo contains a display, opened in 2016, called Land of the Lions, based upon the lions who live near the village of Sason Gir in the state of Gujarat, India. It contains replicas of a Gujarat fortification, train station, bungalows, rickshaws, bicycles, carts, and market stalls (staffed by Indian-looking people), plus four lions in outdoor enclosures. Inaugurated in 2016 by Queen Elizabeth, the exhibit revels in imperial nostalgia, recalling the zoo's founding president Stamford Raffles (1826), who ruled Java and established the British colony of Singapore. The zoo-village erases from history more than a century of resistance to direct British rule, led at the end of the period by the Gujarat-born, Mahatma Gandhi. In the meanwhile, the wild lions of Gujarat are dying off at an alarming rate—184 in the last two years alone from a population of just under six hundred. The reasons are disease, poaching, and accidents (falling into wells, being hit by trains), and the failure of state and federal governments—or international donors—to provide funds to expand the lion reserve and protect it from surrounding development.

In addition to their role in affirming colonialism, nineteenth-century zoos were places of spectacle: They provided arresting images—of animals and other people—that attracted and engaged large and diverse urban crowds. Journalists and travel writers frequently remarked that zoological gardens were places where social classes intermingled in relative harmony. This was no small thing in a century punctuated by social revolutions and the rise of working-class political organizations. At the zoo, visitors of all ranks shared a certain camaraderie, or at the very least, were free to gaze upon one another with either condescension or resentment. Meandering among exotic trees and plants suggestive of distant colonies, men and women of all ages and backgrounds could enter Oriental-looking or otherwise impressive zoo buildings, look at the animals, and be affirmed in their divinely ordained superiority. The capitalist zoo promoted the ideologies of speciesism and human exceptionalism to all visitors, regardless of social class. Whatever else a visitor may be, he or she was not an animal, and even the less advantaged among them could demand the right not to be treated like one. When I was working with Illinois prisoners' families a few years ago to protest a notorious supermax (solitary confinement) prison, the complaint I constantly heard was "the men in prison are treated like animals in a zoo." The point being that humans may be distinguished from animals by their right not to be locked up in small cages and abused.

Most animals in nineteenth-century zoos would have been either listless or agitated, the result of poor health, boredom, and stress. Many performed stereotypic behavior—a version of human Obsessive-Compulsive or Non-Suicidal Self-Injury disorders—arising from long-term close-confinement: pacing, fur plucking, rocking, swimming in circles, and scratching to the point of self-injury. And it took an enormous toll. Even by the end of the nineteenth century, it was rare for primates at most zoos to live more than two or three years. Mammals fared a little better, but only a very few lived a normal lifespan. In the course of the next century, better veterinary care improved the longevity of the majority of zoo animals, but physical and psychological illnesses nevertheless plagued them, and still do to this day. Female African elephants have a life span of fifty-six years in Amboseli National Park, but only seventeen in zoos. From 2007 to 2009, thirty-nine of fifty-seven Asian lion cubs in European zoos—including those in London—died before reaching four months of age. Between 2000 and 2014, the chance of a lion surviving to reproductive age in a European zoo was only 44 percent.

Until his death in 2011, the polar bear Gus at the Central Park Zoo was a special attraction because of his compulsive figure-eight swimming. Few zoo visitors understood this was an illness. Recent studies of captive polar bears indicate that 85 percent of them spend at least one-quarter of their days engaged in stereotyped movement. It's not surprising; their enclosures are only one-millionth the size of their natural range. At the Kovlar Lion House at the Lincoln Park Zoo in Chicago, where I recently visited, a solitary lion could be seen pacing back and forth in her cage. When I asked a keeper about this stereotypic behavior, she replied: "The lion also has access to an outdoor area, but it's cold today. It is very difficult to keep animals stimulated in the limited space we have." Sue Coe observed and recorded the stereotypic behavior of tigers at the zoo in Grand Rapids, Michigan [PAGE 99]. They are a disorienting blur, like a Futurist painting by Giacomo Balla.

Some captive Killer Whales or Orcas, such as Morgan, currently held at Loro Parque in Spain's Canary Islands, also suffer from OCD and self-injury disorder. Found separated from her pod and in danger, she was rescued in 2010 off the coast of Norway as part of a "rehabilitation and release program," but a year later, was nevertheless sent to Loro and made to perform before audiences. There she was regularly attacked by other distressed Orcas, and in 2016 was seen repeatedly slamming her head against a steel gate, grinding her teeth on concrete, and even beaching herself on the side of her pool. The latter behavior is potentially suicidal since Orcas will die if they remain out of water for more than a short time. It is not clear whether she beached herself to escape the bullying of other whales or out of despair at her captivity, but either way, it's a manifestation of the suicidal behavior sometimes found among zoo animals.

Until recently, researchers dismissed as mere anthropomorphism the proposition that animals attempted or committed suicide, what we are calling zooicide. The argument was that suicide required both self-consciousness and an understanding of death, and that animals lacked both. (Some also argue that suicide requires possession of "free will," however that's a theological argument beyond rational discussion.) In fact, lots of animals have demonstrated a consciousness of self: the great apes, dolphins, elephants, magpies, crows, and others. (The usual test for self-consciousness, which involves looking into a mirror, is imperfect. Some animals simply lack the physiology to react to their own mirror images. Self-consciousness of some kind may be characteristic of nearly all animals.) In addition, many animals recognize death. Elephants, chimpanzees, and lots of other animals perform death rituals or visibly grieve, as observers since antiquity have noted. Dogs are so affected by the loss of a companion that veterinarians increasingly counsel that they be allowed to inspect the body of their departed friends so they can better process the death and achieve closure. Regardless, it is not at all clear that an understanding of death is a necessary precondition of suicide. Many humans lack a conception of death, for example children, and yet commit suicide anyway. When Morgan climbed up onto a concrete platform, effectively beaching herself, she was engaging in self-injurious behavior seen countless times among animals confined in zoos. It was a failed zooicide.

Another famous Orca, Tilikum, died in 2017 after thirty-three years in captivity, the last twenty-one at SeaWorld in Orlando, Florida. Tilikum experienced enormous stress during his years in aquaria and, as a result, was involved in the killing of three keepers. Attacks on humans by wild Orcas are almost unknown, and Tilikum may have expected to be killed as punishment for his violence, a version of "suicide by cop." But that's not what happened. Under pressure from the public and stockholders, as well as negative publicity generated by the documentary *Blackfish*, SeaWorld agreed in 2016 to phase out killer-whale performances and stop breeding the animals, though they remain in captivity. And Orcas are not the only cetaceans that suffer in aquaria. Beluga whales, dolphins, and porpoises similarly languish. The death of two Belugas just weeks apart in the Vancouver Aquarium in 2016, and the resulting public outcry, ultimately led to a decision by the aquarium to end the practice of holding cetaceans in captivity. Tilikum died of illness at SeaWorld in 2017.

In addition to death from injury and physical or psychiatric disease, thousands of animals in European and American zoos

The Capitalist Zoo

and aquaria are killed every year because they are determined to be redundant, dangerous, poor candidates for breeding, or simply in surplus of what is needed for exhibition. In Europe, according to the European Association of Zoos and Aquaria, this program of "management euthanasia," kills between 3,000 and 5,000 animals a year. The total in the U.S. is unknown, but it has been estimated that 45 percent of zoos regularly cull their populations. Some of this is because of captive breeding. The usual justification for this practice is greater genetic diversity and species protection, according to the American Association of Zoos and Aquaria (AZA) approved Species Survival Plan (SSP). But since almost no zoo animals are ever returned to the wild, the argument is unpersuasive. In most cases, animals are bred because their young are popular with zoo visitors. Captive breeding however, besides the frequent cruelty of the process itself, creates an overpopulation leading to the transfer, exchange, and sale of animals—separating friends, mates, and mothers and babies—as well as the killing of adults. The consequence of the practice became clear to many people in the case of the polar bears Szenja and Snowflake, separated by SeaWorld of San Diego after twenty years, so that Snowflake could be used to breed more polar bears at the Pittsburgh Zoo. Two months after their separation, Senja died, apparently of a broken heart.

Zoo officials have long recognized the psychological impact upon animals of close captivity. The first significant efforts to alleviate it occurred at the privately owned Hamburg Zoo, or Tierpark, under the direction of Carl Hagenbeck at the turn of the twentieth century. Hagenbeck, otherwise notorious for his brutal transport and trade of animals, introduced more naturalistic and bar-free enclosures "to give the animals the maximum of liberty," as he wrote. He sometimes placed different species in the same enclosures, or at least simulated proximity, while discretely separating animals through the use of rock walls and hidden moats. But his efforts to provide liberty were superficial and in fact had only a modest impact on zoos in subsequent decades. In 1900, the superintendent at Woodward's Gardens (a private zoo and amusement park in the Mission District of San Francisco), told a researcher that "several species commonly end their cage lives in lunacy.… Captive bears are apt to fall into a sort of sullen despondency. Foxes and cats often go crazy.… The higher apes and baboons rarely thrive in cages. Soon or late they become abnormally vicious or have a complete, physical breakdown." Monkeys at the Paris Zoo generally survived about a year and a half.

Even today, primates often fare poorly in zoos, and efforts to improve their lives through so-called "enrichments," are inconsistent at best. Gorilla enrichment includes both "social" and "physical" components, according to the AZA approved SSP: "Examples of social enrichment include housing with other gorillas and providing opportunities for interaction with keepers. Examples of physical enrichment include novel presentation of food items, variation in living spaces, and provision of objects for play or interaction." Actual enrichment activities for gorillas and other primates are often quite meager, even at the best zoos. They may consist of little more than placing food in unusual containers and the provision of ropes, plastic bones, or other toys, as well as TVs and mirrors. At the Western Lowland Gorilla exhibition at the Regenstein Center for African Apes in the Lincoln Park Zoo, the animals are kept behind a glass wall in an enclosure that totals 29,000 square feet, inside and out. The space for the gorillas however, is much smaller since the facility is shared with the chimpanzees. (In the wild, gorillas travel in ranges from eight to thirty square kilometers.) During one visit, I spoke briefly to a keeper who I saw reaching into an enclosure, offering up an iPad. If the gorilla hit the right color bars in the right order, he got a treat. The researcher described this as enrichment.

At the Bronx Zoo's Congo Gorilla Forest, enrichment, as well as space, is also highly limited. Gorillas, mandrills, colobus monkeys, Red River Hogs, and okapi appear to roam freely, but each is actually confined to a small part of the 6 ½ acre enclosure. The naturalism of the habitat is created for the benefit of zoo visitors, not animals. At Zoo Atlanta, which contains the largest population of Western Lowland Gorillas in North America, the animals are contained in a 1.8-acre habitat, subdivided by double moats into five smaller areas. According to the zoo website, gorillas have the opportunity to jump in piles of leaves. In addition, keepers employ "retired fire hoses…an almost indestructible material that can be formed into hammocks, swings, and such." Gorillas may also rummage through piles of dead leaves to find fruit and edible salad greens that are scattered every day. Do the gorillas know that the food has been purposely scattered for their benefit? Do they ask themselves why their keepers won't just give them the food directly?

In London, according to a sign outside their enclosure, Gorillas receive "enrichment" from "Hessian sacks filled with something smelly," "ice-lollies in the summer to keep them cool," "hiding food in boxes of straw," and "bottles of fruit tea." (The patronizing language is excruciating.) Other primates have similarly limited opportunities for enrichment. During my last visit, I saw a custodian walk past the screened enclosure for the White-naped Mangabey, hurling handfuls of shelled peanuts for the animals to find and eat. The scavenger hunt kept the animals busy for about ten minutes. In the wild, monkeys, like gorillas, forage up to fourteen hours a day.

But in fact, animal care in zoos—particularly care for the psychological health of animals—remains poor. Zoo installations, as Detroit Zoo director Ron Kagan has noted, "attend disproportionately to human needs and wishes," not to animal welfare. And enrichment programs are neither universal nor implemented with consistency. Jenny Gray, president of the World Association of Zoos and Aquaria, has written that only about 3 percent of the 10,000 zoos worldwide provide their animals with adequate welfare, by which she means proper veterinary care as well as enrichment. Most animals continue to suffer from small enclosures, limited time outside, and little interaction with friends or family members, as ethologist Marc Bekoff has discussed. The reasons for poor animal welfare are lack of time, limited space, and poor training—and in some cases indifference. Another is that zoo custodians routinely move between exhibits, regardless of their animal expertise, if any. At most zoos, in other words, polar bear keepers can get transferred to the cockatoo cage, and crocodile minders shifted to the penguin house, regardless of their knowledge of the ethology.

In 1964, agricultural officials in the U.K. proposed what they called the "Five Freedoms" to improve the welfare of domesticated animals. These are: 1) freedom from injury and disease; 2) freedom from hunger and thirst; 3) freedom from thermal or physical distress; 4) freedom to express most normal behaviors; and 5) freedom from fear. Though none of these in fact became the norm in animal agriculture in the U.K., the U.S., or elsewhere, a number of zoo authorities have embraced them and added two more: freedom for the animal to direct the quality of its life, and freedom from boredom. The passage in 1970 of the U.S. Animal Welfare Act was similarly intended to improve the circumstances of both domesticated and wild animals in captivity.

But without autonomy, these enumerated freedoms are sops. In fact, zookeepers assume complete control of their charges. They feed and house the animals, provide (or don't provide) companions or mates, control reproduction, and transfer or destroy animals based upon the needs of the institution and the SSP. Zoo animals, like domesticated animals, are thus little more than chattels who may be used, abused, or discarded as needed. Upon arrival from the wild or from other zoos, or from birth, they are subject to *thingification*, which Césaire used as a synonym for colonization, or the turning of a person into "an instrument of production." Zoo animals serve colonial ideology, zoo administrators, corporate sponsors, and just as importantly, animal agriculture.

The Zoo and Capitalist Agriculture

The capitalist zoo arose and developed along with capitalist agriculture. They are two sides of the same coin. As the pace of national and international trade and industry accelerated in late-eighteenth-century England, and subsequently the rest of Europe, North America, and elsewhere, existing animal agriculture did not provide the profits available in other industrial sectors, such as mining, textiles, and grain production. It was thus essential that it be rationalized and made more productive. The development of coal-based steam power in the mid-nineteenth century offered the possibility of shipping live animals large distances across land and sea, and turning regional and national markets into more profitable international ones. Refrigerated railway-cars, developed in the 1860s, facilitated the expansion by permitting packaged meat to be transported without spoilage. Stockyards, such as the Union Stockyards in Chicago, and adjacent slaughterhouses were soon organized as modern factories, and by the early-twentieth century, became the model for the assembly line used by Henry Ford in the production of his Model T car at the River Rouge Plant in Detroit.

Each step in the development of capitalist agriculture and production was accompanied by a widening of the moral divide between humans and animals. The more animals were segregated from humans, concentrated together in feedlots, packaged and turned into products bearing little resemblance to living beings, the more people came to accept the thingification of animals—their transformation into instruments of production. Whatever association domesticated animals may once have had with the countryside, rural life, and face-to-face society disappeared as meat became available in large quantities in butcher shops and then supermarkets and fast food outlets. Whatever symbolism wild animals had—the lion, elephant, bear, hyena, and ostrich—was erased as they were concentrated in zoos and became a mere genus-species associated with a particular geography, overseas colony, or habitat. In 1848, Karl Marx, writing from London, described how modern capitalism removed the "halo" from "every occupation hitherto honored and looked up to with reverent awe. It has converted the physician, the lawyer, the priest, the poet, the man of science, into its paid wage laborers." The same may be said about animals. The capitalist zoo stripped of its halo every animal hitherto honored and looked up to with reverent awe. It converted the elephant and the tiger into menagerie and zoo attractions. It changed the orca, dolphin, and seal into acrobats, paid with wages of dead fish.

The close bond between human and non-human animals—forged in the pre-historic, communal age and injured by tributary society—was thus nearly destroyed by the growth of capitalism. The change wasn't immediately apparent in the nineteenth and twentieth centuries and even now is hard for many people to acknowledge. When we observe a flock of sheep or a herd of cattle grazing on a hillside or in a field, we may suppose that the human relationship with animals is as direct as it was during the tributary age: the animal grazes today so that people may eat a lamb chop or a hamburger tomorrow. In other words, animals are killed so that humans may consume their meat. In fact, however, the relationship between animal and consumer is highly mediated by the process of exchange. Animal agriculture treats animals as raw materials that, when combined with human labor power and factory production, creates a commodity with a value considerably greater than the original animals, as summarized by Marx's general formula for capital: M-C-M' (Money is transferred into a commodity which is then sold for more money than before. In more advanced forms of capital, labor power and technology combine with raw material to substantially increase value.) That surplus value or profit (M'), in turn, permits acquisition of more animals, the purchase of more human labor power and factory production, and the generation of still more profit in a cycle without end. At no point in the production cycle therefore, is the animal truly an animal at all, any more than a slaughterhouse employee is a person, or a grazing area part of nature; all are simply "instruments of production," or inputs, deployed in the manner that is most efficient and profitable for the business that owns or controls them. To be sure, use-value, the consumption of meat, is a necessary prior condition for animal agriculture. But it is not what accounts for its vast size, methods of production, or profitability.

The functioning of zoos and the purpose of zoo animals in the capitalist age is also mostly opaque to the casual observer. Zoos present themselves as scientific, educational, and recreational facilities, and their animals as cherished guests. But in fact, they perform generally un-acknowledged economic functions that belie any scientific, educational, or conservation rationale. To begin with, zoos are businesses. If they are private (and there are at least ten times as many private as public zoos in the U.S.), they generate profits. In 2015, the San Diego Zoo had assets of $545 million and finished the year with a surplus of $30 million. During the previous five years, it collected about $175 million more than it spent. The San Diego Zoo also generates profit in other sectors, especially tourism, including hotels, restaurants, and myriad concessions. According to a study from 2012, it generated some $840 million annually in local economic activity. Even small zoos with no research function at all may be economic engines for a community.

Last year, there were around 180 million visits to U.S. zoos and aquaria, more than the combined attendance for all major league sports. Admission fees range from about $20 to $50 per person. (Children are charged $40 at the San Diego Zoo.) But admission fees don't cover all the costs or generate all the profits. The biggest zoo sponsors in the U.S. are Coca-Cola, PepsiCo, Bank of America, Wells Fargo, Blue Cross and Blue Shield, Mattel, Walmart, AT&T, PNC Bank, and JP Morgan Chase, though zoo sponsorships differ from region to region. At the Bronx Zoo, sponsors include BP, Chevron, Citibank, Goldman Sachs, and Dow Chemical. These sponsorships bring corporations significant returns in the form of general good will, new investments, community prestige, and a favorable regulatory environment.

The capitalist zoo finally, contrary to claims, contributes little to species preservation or public education. Only about 1 percent of AZA accredited zoo budgets goes to conservation, an average of about 89¢ per visitor. A vanishingly small number of zoo animals are ever returned to the wild, and indeed, the great majority of zoo species are not endangered. Few zoo visitors, recent research has shown, learn more than the most rudimentary facts about animal lives, global warming, and the destruction of habitat. (That will surprise nobody who has recently been to a zoo.) Nor are visitors instructed about the necessity of veganism to reduce global warming and protect habitats. Indeed, the food concessions at major zoos, including McDonald's, Burger King, and Taco Bell, impart the opposite lesson. And even though the didactics at zoos are primarily addressed to the youngest visitors, many children, according to research from 2011, actually register a more negative attitude toward zoo animals after their visits than before. Rather than intervening to slow habitat destruction, global warming, poaching, overfishing, and the illegal trade in exotic animals or their parts, zoos are mostly passive witnesses to the crash of wild animal populations.

Zoos enable animal agriculture and vice versa—they are two sides of the same coin of exploitation. By so clearly distinguishing exotic from farm animals, the capitalist zoo proposes that while the first deserve protection, the second are fated for slaughter. However, the opposite idea is also promoted: by carefully policing the boundary between domesticated and zoo animals, capitalist agriculture asserts that while domestic animals are familiar and necessary, the zoo animals are exotic and remote. They are valuable for entertainment but their existence or non-existence is ultimately of little significance. Thus, zoos and animal agriculture together enable people to cherish some animals, while torturing and killing many others; but because of the interaction I have described, both wild and domestic animals are destroyed: the catastrophe that has befallen the one has descended upon the other. And so, the solution to the crisis of animal survival is inescapable: break the bond between animal agriculture and zoos. Boycott zoos or go vegan—even better, do both.

Darkness and Light

Sue Coe is an accomplished painter, but the majority of her works are drawings or prints that rely upon the drama created by darkness and light. Sunshine and shadow, deep gloom and sudden illumination, and half-light—that in-between that can trigger feelings of anxiety or even terror—comprise Coe's principle formal language. It's there in her early works such as *Sharpeville* ([BELOW], 1982), from the anti-Apartheid collection published as *How to Commit Suicide in South Africa*; *Wheel of Fortune* (1989), which depicts a pig slaughterhouse like a Dantean circle of hell; and her *AIDS Suite* (1994), which unflinchingly depicts the suffering of victims while implicitly condemning the U.S. government's slow response to the epidemic. Coe's best-known single work may be the drawing *Modern Man Followed by the Ghosts of His Meat* ([PAGE 17, TOP], 1990), later produced as a large edition photo-etching. It's a nighttime vision of innocent animals haunting a fast-food eater. It's one reason I became a vegan.

In deploying the resources of black and white, Coe rightly places herself in the lineage of graphic artists that include Albrecht Dürer, Rembrandt, Francisco Goya, Théodore Géricault, Honoré Daumier, Odilon Redon, and a number of German artists from the early-twentieth century including Otto Dix, Käthe Kollwitz, and George Grosz, as well as the

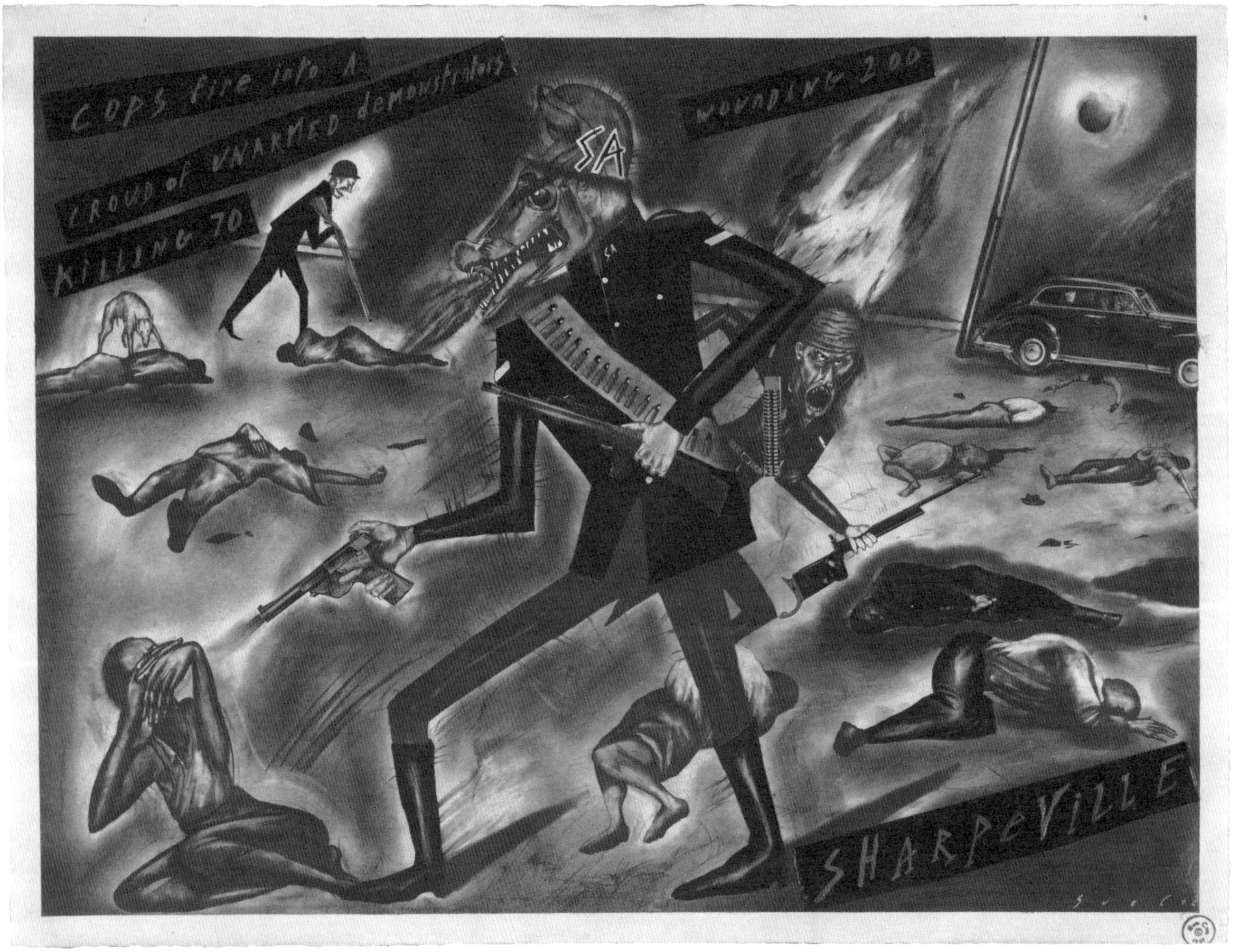
COPS fire into A
CROWD of UNARMED demonstrators
KILLING 70
WOUNDING 200
SA
SHARPEVILLE

Sue Coe, *Sharpeville*, 1982.

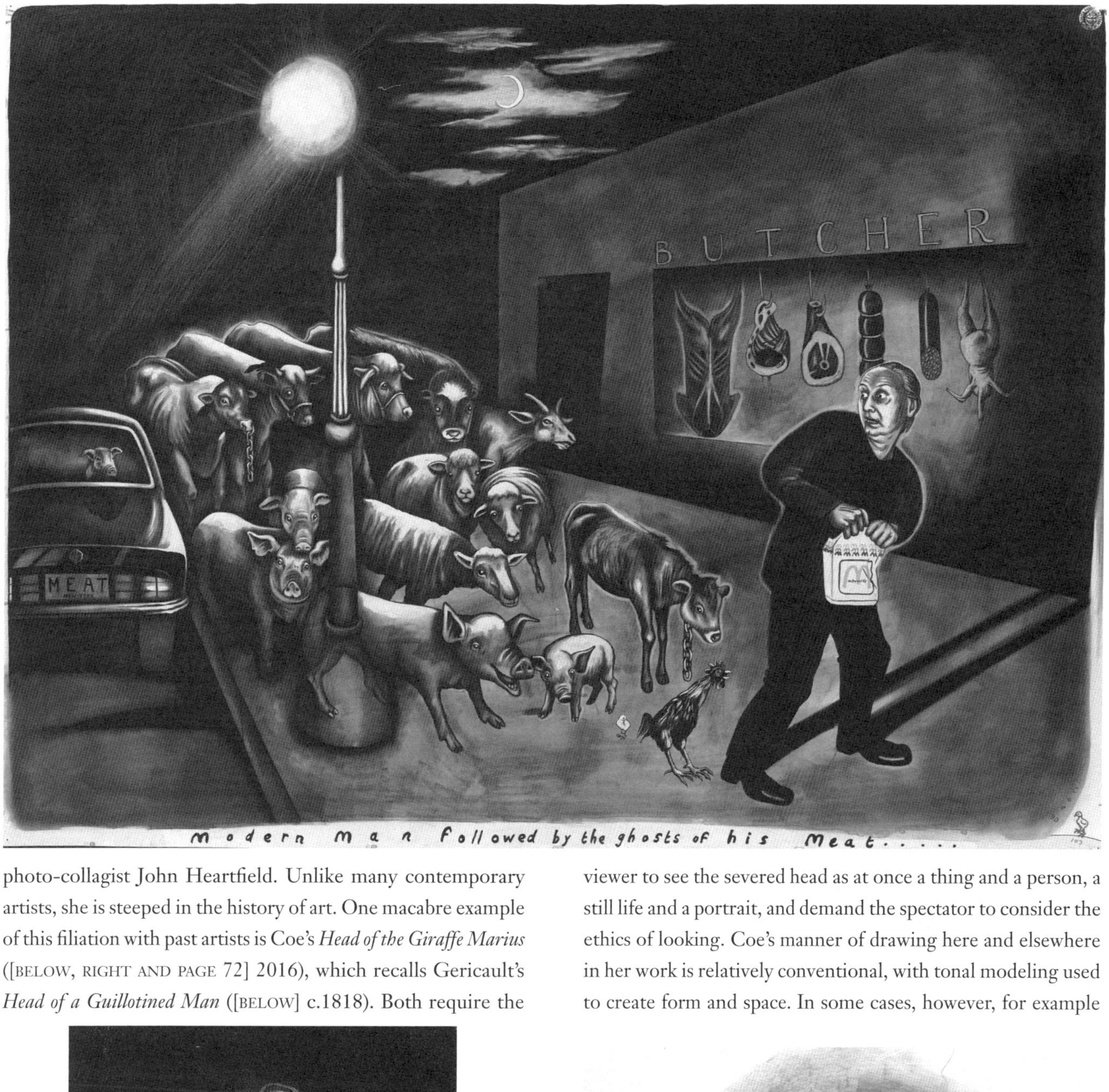
BUTCHER
MEAT
modern man followed by the ghosts of his meat....

Sue Coe, *Modern Man Followed by the Ghosts of His Meat*, 1990.

photo-collagist John Heartfield. Unlike many contemporary artists, she is steeped in the history of art. One macabre example of this filiation with past artists is Coe's *Head of the Giraffe Marius* ([BELOW, RIGHT AND PAGE 72] 2016), which recalls Gericault's *Head of a Guillotined Man* ([BELOW] c.1818). Both require the viewer to see the severed head as at once a thing and a person, a still life and a portrait, and demand the spectator to consider the ethics of looking. Coe's manner of drawing here and elsewhere in her work is relatively conventional, with tonal modeling used to create form and space. In some cases, however, for example

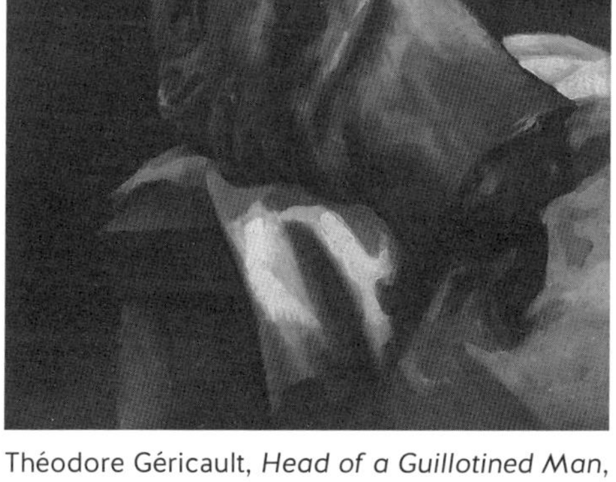

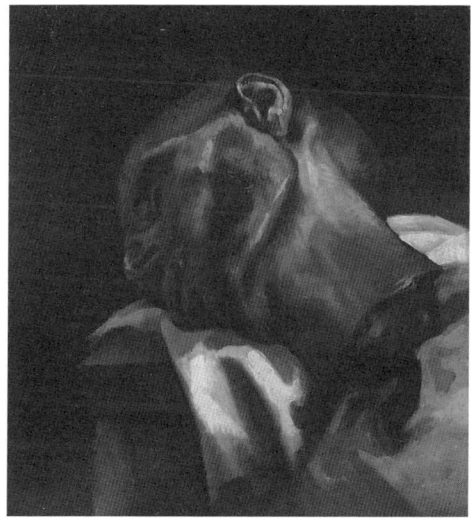

Théodore Géricault, *Head of a Guillotined Man*, 1818.

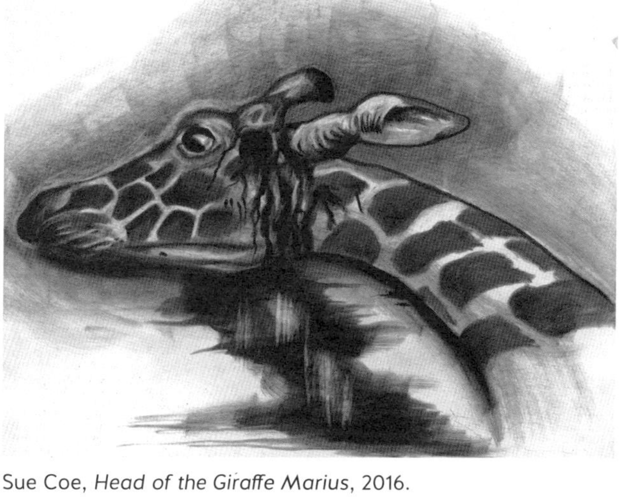

Sue Coe, *Head of the Giraffe Marius*, 2016.

the drawings titled *Butcher* ([BELOW] 2011) and *Never on Her Knees*, the artist utilizes both the heightened chiaroscuro and flattened or telescoped space of the Expressionist painter and graphic artist Max Beckmann to recall the Nazi era and its multitude of crimes. The inevitable conclusion is that animal slaughterers are like Nazis and meat eaters are their willing accomplices. Fascism is found, she argues, wherever the powerful exult over those with less strength, wealth, or opportunity.

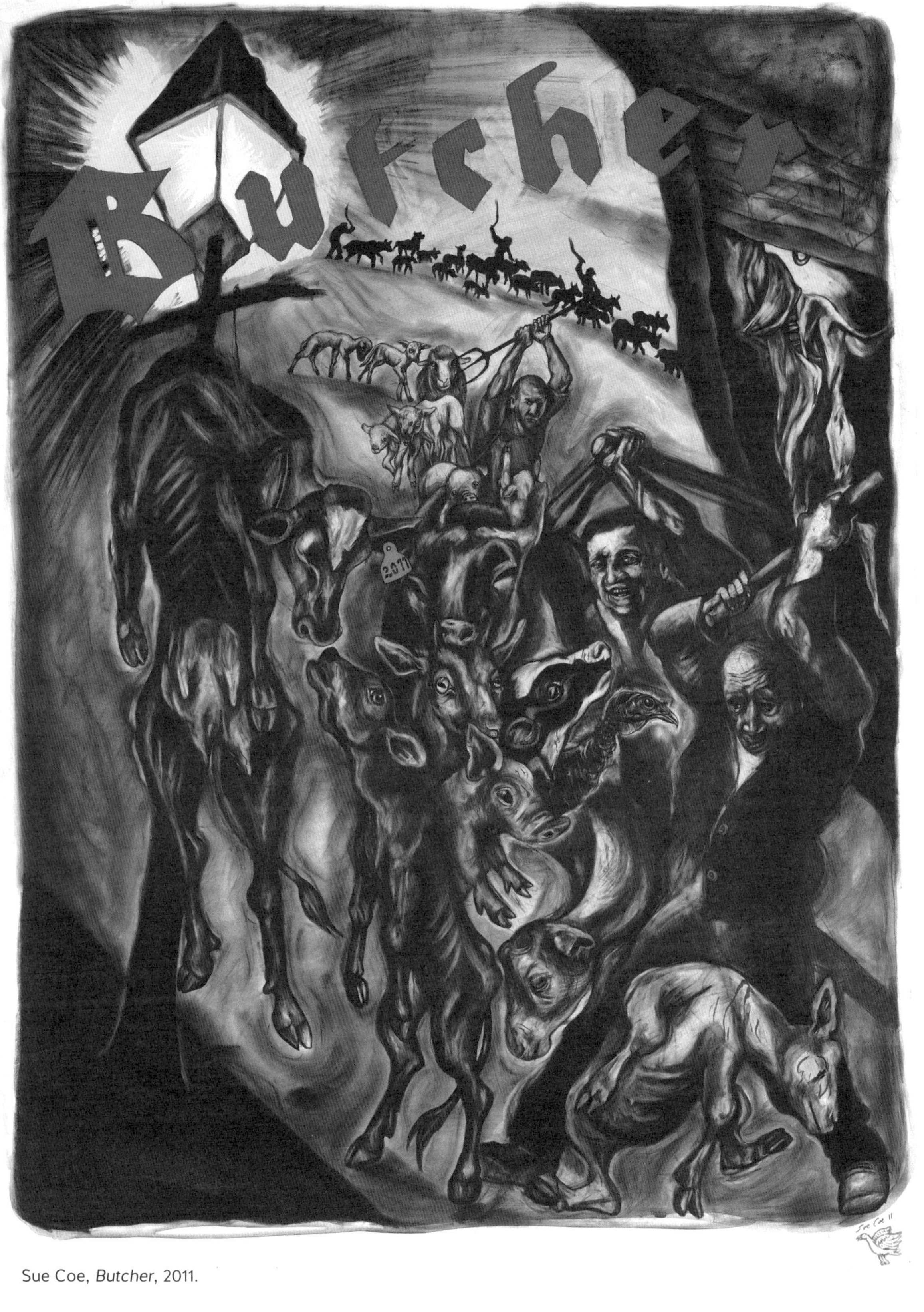
Butcher

Sue Coe, *Butcher*, 2011.

The drama of black and white is everywhere apparent in *Zooicide*. It's found in the *Harambe* [PAGE 36] and *Szenja* [PAGE 76] linocut series, the tragic narratives of which unfold like a mini-graphic novel. We see it in the grotesque lithograph *Conservation* [PAGE 29], which depicts the Grim Reaper feeding exotic animals—a bear, giraffe, and elephant—into the gaping maw of a human zoo. It's also apparent in a pair of poignant giraffe drawings, *Painted Friend* and *Mother and Child* [PAGES 70 AND 71]. At a purely formal level, both drawings rely upon a middle tone of gray—used as a background—to bring the animal protagonists into sharp relief. Their patterned bodies are themselves, of course, comprised of the contest of black and white. The first of these works is also a subversive meditation upon the power of art. The ancient Greek painter Zeuxis, according to Pliny the Elder, was so skillful at mimesis that when he painted a still life with grapes, birds alighted and attempted to eat them. Coe shows something similar—a giraffe lowering her neck to nuzzle the form of an antelope painted on the wall of her enclosure. But what the artwork really argues is that art can sometimes be a handmaid to cruelty. The keepers at the Como Zoo are so heedless of the needs of the animals in their care, that they will encourage an animal to seek comfort through physical contact with cold bricks painted to resemble a living being. *Mother and Child* is thus the rebuke to *Painted Friend*, a reminder that love is found wherever animals are, and that art is not a substitute for a living thing.

Circus Train Wreck ([BELOW AND PAGE 25] 2010) is a big drawing that relies upon subtle light effects to heighten a surreality found in the subject itself. In May 1893, a train carrying animals belonging to the Walter L. Main Circus and Menagerie crashed in Tyrone, PA. Fourteen out of seventeen rail cars tumbled down a ravine, either killing or freeing the captive animals. The dead included fifty horses and a pair of cows described by circus promoters as "sacred." Among the liberated were a gorilla, tigers, lions, zebras, alligators, elephants, and camels. Many of the escaped animals—including the gorilla and elephants—were

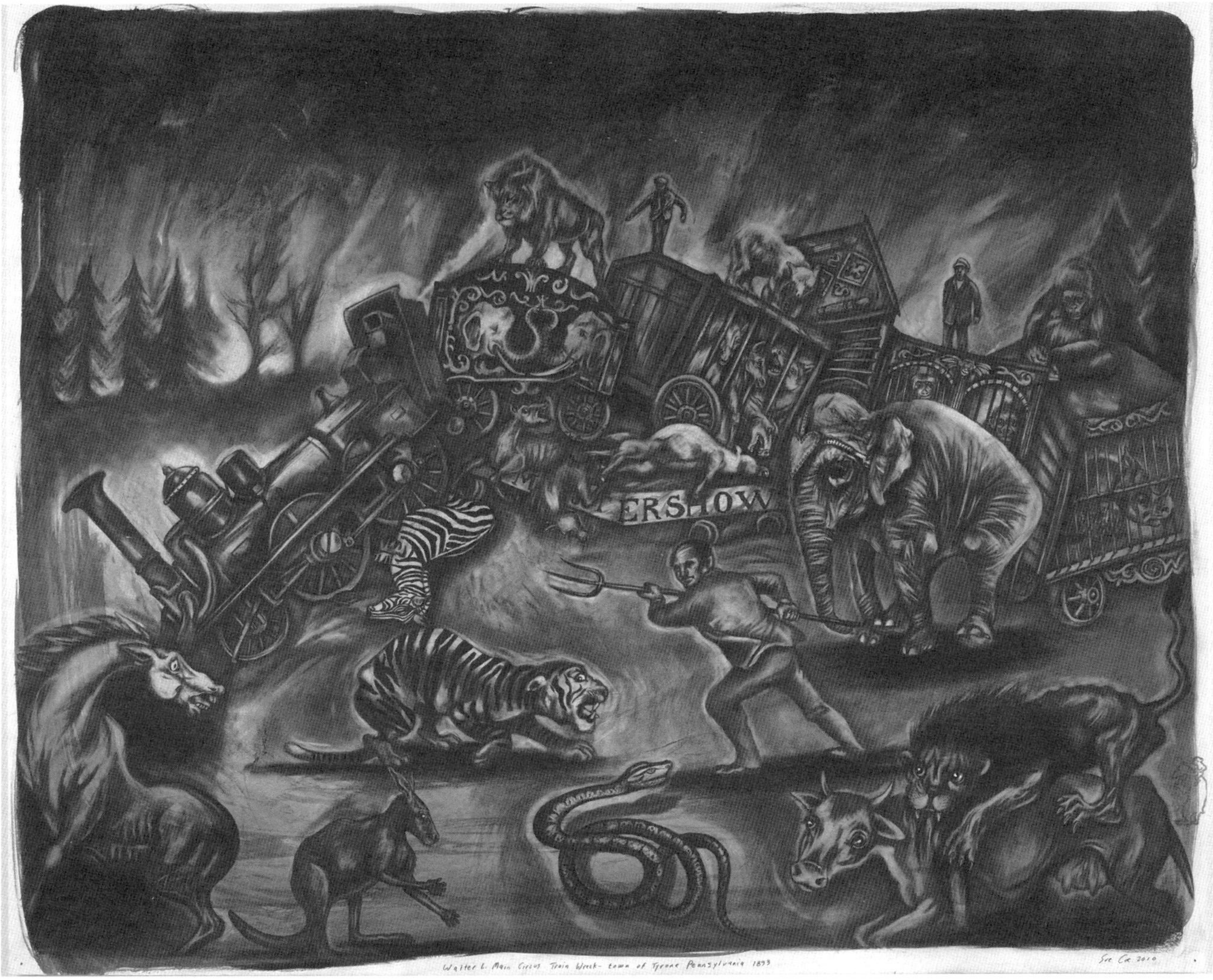
ERSHOW
Walter L. Main Circus Train Wreck- town of Tyrone Pennsylvania 1893
Sue Coe 2010

Sue Coe, *Circus Train Wreck*, 2010.

George Stubbs, *Horse Attacked by a Lion*, 1769.

quickly recovered, but others were killed, including a tiger who remained free for several days but was shot to death after attacking a local (not holy) cow. Some animals, including a kangaroo and a few snakes, appear to have escaped into the nearby woods and were never recovered. Two years later, someone claimed to see a kangaroo hopping across a road, but the sighting was never confirmed, and there are no known marsupials (apart from opossums) roaming today in eastern Pennsylvania.

Coe's drawing of the scene collapses several days into a moment—a crepuscular snapshot of destruction, death, fear, and perhaps hope. The composition is a great, flattened pyramid anchored at the bottom by a frightened horse at lower-left, a kangaroo, a snake, and a lion attacking a cow. Above that we see a man with pitchfork battling a tiger, as well as a dazed elephant, a dead zebra and five mangled railway cars, their animal cargo either free or trapped inside. (A rhinoceros is visible in the carriage behind the elephant.) At the upper right, a gorilla—he was known as The Man-Slayer—sits placidly on top of a train carriage surveying the scene. As usual in Coe's works, the history of art is present in abundance: the frightened horse recalls the one attacked by a lion in George Stubbs great painting of a *Horse Attacked by a Lion* ([ABOVE] 1769); the coiled snake invokes the famous Gadsden flag (1775) from the American Revolutionary War that contained the motto: "Don't tread on me"; and the lion attacking the cow again reminds us of Stubbs's painting, and also the well-known Hellenistic sculpture of a *Lion Attacking a Horse* (c. 300 BCE) in the Capitoline Museum in Rome.

Beyond that, Coe has created a kind of inversion of the Quaker Edward Hicks's famous paintings (he made over a hundred of them) titled, *The Peaceable Kingdom* [BELOW]. These naively rendered landscape and animal pictures illustrated the passage from Isaiah (11:6): "The wolf also shall dwell with the lamb, and the leopard shall lie down with the kid, and the calf and the young lion and fatling together; and a little child shall lead them." Coe depicts the earthly domain of humans and

Edward Hicks, *The Peaceable Kingdom*, 1834.

animals as anything but peaceable in *Circus Train Wreck*, but the temporary emancipation of some of the animals foretells an eventual animal revolution, a subject Coe treated explicitly in her *The Animals' Vegan Manifesto* (2017). In several images from that book and elsewhere in Coe's work, the dawn of animal liberation is figured in the form of a sunrise, a motif frequent in anarchist, socialist, and utopian literature and art of the late-nineteenth and early-twentieth centuries. That same emancipation is suggested in *Circus* too, in the light that silhouettes the trees in the background.

The motif of the rising sun is also found in *Faithful Elephants* [PAGES 61–68], a painting that recounts the story of the Ueno Zoo in Tokyo. Here the red sun represents the Japanese flag, while also perhaps signaling a future time of redemption. In this repeated recourse to a rising sun, Coe offers viewers a temporal as well as political perspective. *Circus Train Wreck*, *Faithful Elephants*, and many other works are situated in the future as well as the past, after the animal revolution. This engagement with "looking backward" is also implicit in *Zooicide*, for example in a poignant linocut showing the artist herself at work. *Reflection* ([BELOW, LEFT] 2018) depicts Coe sketching in a notebook, her face and torso reflected in the glass of a tank that holds an octopus. Several starfish float past, as if they were celestial bodies unfixed from the celestial firmament.

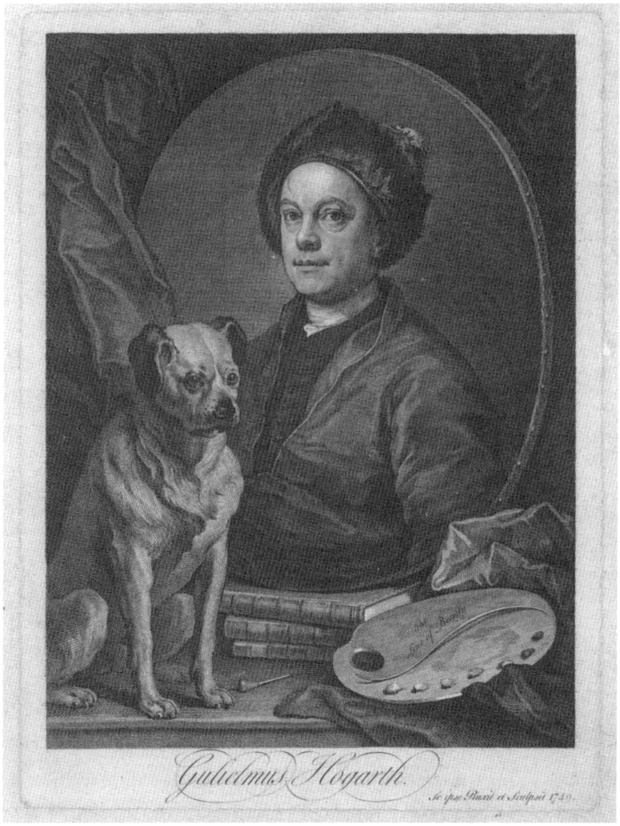
Gulielmus Hogarth.
William Hogarth, *Gulielmus Hogarth*, 1748–1749.

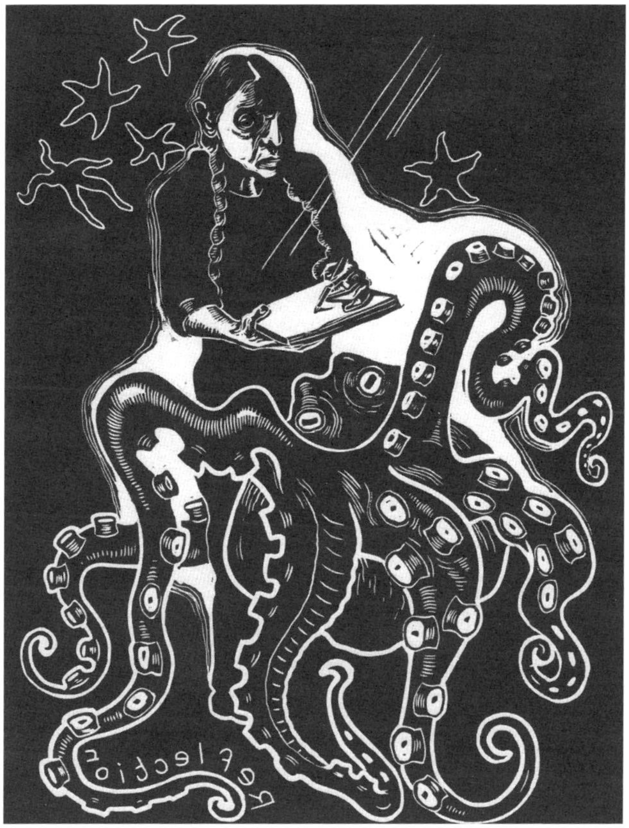
Sue Coe, *Reflection*, 2018.

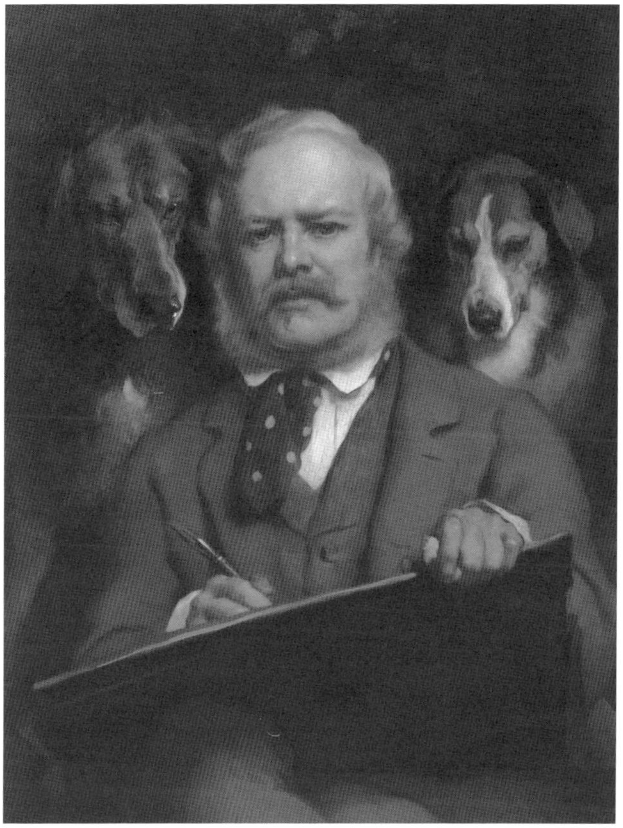
Edwin Landseer, *Portrait of the Artist with Two Dogs*, 1865.

Coe's self-portrait linocut recalls a number of prior self-portraits-with-animals in the history of art, including Hogarth's *Gulielmus Hogarth* ([PAGE 21, TOP RIGHT] 1749) showing the artist and his pug named Trump, Edwin Landseer's *Portrait of the Artist with Two Dogs* ([PAGE 21, BOTTOM RIGHT] 1865), and Frida Kahlo's *Self-Portrait with Monkey* (1938). Each of these earlier works reveals a strong identification between artist and animal: in the first case a common feistiness (pugnaciousness); in the second an acknowledgement of the shared nature of the professional enterprise; and in the third an equivalent wildness and hirsuteness. But how could Sue Coe have identified with an eight-armed mollusk from the order Octopoda? Coe recently told me: "She was in a tank at Syracuse Zoo, and I loved her more than anything. She shows stereotypical behavior."

Octopuses, we now know, are extraordinarily intelligent despite the fact that they have an entirely different body-form than humans, mammals, or birds. (Two-thirds of their neurons are housed in their tentacles, meaning that they think with their fingers as well as their head.) They have great memories (both long and short term), remarkable problem-solving abilities, the ability to make and use tools, and a penchant for playfulness. Most of all, they love freedom and will perform extraordinary feats in order to escape any container that holds them. And this may be why Coe is so attracted to octopuses; they express her overall, political project with utmost clarity: 1) to recognize the personhood of all other individual animals, even those evolutionarily most remote from us; 2) to acknowledge, through the act of witnessing and reporting, the suffering of animals; 3) to protest the transformation of animals into capitalist commodities, and their *thingification*; and finally, 4) to represent the agency of animals, that is, their will to freedom, and their desire to maintain lives in keeping to their own natures. *Reflection* announces that all animals held in captivity or otherwise denied their freedom—non-human and human alike—are victims of zooicide, and that like Sue, we all have the responsibility to witness, represent, and resist.

—Stephen F. Eisenman

Sources:

Almiron, Nuria. "Slaves to Entertainment: Manufacturing Consent for Orcas in Captivity." In *Animal Oppression and Capitalism*, edited by David Nibert, editor, 50–70. Santa Barbara: Praeger, 2017.

Amin, Samir. *Eurocentrism*. New York: Monthly Review Press, 2009.

Arnott, John. "From Menageries to Master Plans: Linking the Botanical with the Zoological." *Botanic Gardens Conservation International* 1, no. 2 (October 2004).

Baratay, Eric, and Elisabeth Hardouin-Fugier. *Zoo: A History of Zoological Gardens in the West*. London: Reaktion, 2002.

Barrington-Johnson, J. *The Zoo: The Story of London Zoo*. London: Robert Hale, 2005.

Bekoff, Marc. "Zoo Ethics and the Challenges of Compassionate Conservation: A Comprehensive Interview with Jenny Gray, CEO of Australia's Zoos Victoria." *Psychology Today* (July 18, 2017).

Bekoff, Marc, and Jessica Pierce. *The Animals Agenda: Freedom, Compassion, and Coexistence in the Human Age*. Boston: Beacon Press, 2017.

Berger, John. *Why Look at Animals?* London: Penguin, 2009.

Best, Steven. "Zoos and the End of Nature." http://www.drstevebest.org/ZoosAndTheEnd.htm.

Burns, John. "Sarajevo Journal: In the Zoo's House of Horrors, One Pitiful Bear." *New York Times*, 4, October 16, 1992.

Césaire, Aimé. *Discourse on Colonialism*. New York: Monthly Review Press, 2001.

Christiansen, Keith. "Early Renaissance Narrative Painting in Italy." *The Metropolitan Museum of Art Bulletin* 41, no. 2 (Autumn 1983): 1, 3–48.

Cohen, Simona. "Titian's London *Allegory* and the three beasts of his *selva oscura*." *Renaissance Studies* 14, no. 1 (March 2000): 46–69.

Dierking, Lynn, Kim Burtnyk, Kirsten Buchner, and John H. Falk. "Visitor Learning in Zoos and Aquariums: A Literature Review." Silver Spring, MD: American Zoo and Aquarium Association, 2002.

Donaldson, Sue, and Will Kymlicka. *Zoopolis: A Political Theory of Animal Rights*. Oxford: Oxford University Press, 2011.

Eisenman, Stephen F. "Criticizing Animal Testing, at My Peril." *Altex* 33, no. 1 (February 1, 2016): 3–12.

Eisenman, Stephen F. *The Cry of Nature: Art and the Making of Animal Rights*. London: Reaktion, 2014.

Evans-Pritchard, E.E. *The Nuer: A Description of the Modes of Livelihood and Political Institutions of a Nilotic People*. Oxford: Oxford University Press, 1940.

Falk, John H., et al. "Why Zoos & Aquariums Matter: Assessing the Impact of a Visit to a Zoo or Aquarium." Silver Spring, MD: Association of Zoos and Aquaria, 2007.

Fernandez, Colin, and Glen Keogh. "London Zoo lion family is so inbred that two out of three cubs are dying." *The Daily Mail*, December 27, 2017.

Foster, John Bellamy and Brett Clark, "The Expropriation of Nature," *Monthly Review*, March 1, 2018, p. 2.

Francione, Gary. *Rain without Thunder: The Ideology of the Animal Rights Movement*. Philadelphia: Temple University Press, 1996.

Hargrove, John. *Beneath the Surface: Killer Whales, SeaWorld, and the Truth Beyond* Blackfish. New York: St. Martin's Press, 2015.

Ingold, Timothy, ed. *What is an Animal?* London and New York: Routledge, 1988.

Jamieson, D. "Against Zoos." In *In Defense of Animals: The Second Wave*, edited by Peter Singer. Oxford: Blackwell, 2006, 132–143.

Jensen, Eric. "Evaluating Children's Conservation Biology Learning at the Zoo." *Conservation Biology* 28, no. 4 (March 29, 2011): 1004–11.

Jett, John, and Jeffrey Ventre. "Captive Killer Whale (Orcinus Orca) Survival." *Marine Mammal Science* 31, no. 4 (2015): 1362–77.

Hancocks, David. *A Different Nature: The Paradoxical World of Zoos and the Uncertain Future*. Berkeley: University of California Press, 2001.

Hribal, Jason. *Fear of the Animal Planet: The Hidden History of Animal Resistance*. Oakland: AK Press/CounterPunch, 2010.

Kleiman, Devra G., Katerina V. Thompson, and Charlotte Kirk Baer. *Wild Mammals in Captivity*. Chicago: University of Chicago Press, 1997.

Kramer, Lisa A., and Ray Greek. "Human Stakeholders and the use of Animals in Drug Development." *Business and Society Review* 123, no. 1 (2018): 3–58.

Langa, Mahesh. "Gujarat, Where there is Concern over Disappearing Lions." *The Hindu*, April 7, 2018.

Le Neindre, Pierre, et al. "Animal Consciousness: A Report." *European Food Safety Authority*, 2017.

Litten, Frederick S. "Starving the Elephants: The Slaughter of Animals in Wartime Tokyo's Ueno Zoo." *The Asia-Pacific Journal* 7, Issue 38, no. 3 (September 21, 2009).

Marino, Lori, et al. "Do Zoos and Aquariums Promote Attitude Change in Visitors? A Critical Evaluation of the American Zoo and Aquarium Study." *Society and Animals* 18 (2010): 126–38.

Metz, O., et al. "Retrospective study of mortality in Asiatic lions (Panthera leo persica) in the European breeding population between 2000 and 2014." *Zoo Biology* 36, no. 1 (January 2017): 66–73.

Moss, Andrew, and Maggie Esson. "The Educational Claims of Zoos: Where do We Go from Here?" *Zoo Biology* 32, no. 1 (January–February, 2014): 13–18.

Moss, Andrew, E. Jensen, and M. Gusset. "Evaluating the contribution of zoos and aquariums to Aichi Biodiversity Target 1." Conservation Biology, August 2014.

Neiwert, David. *Of Orcas and Men: What Killer Whales Can Teach Us?* New York: Overlook Press, 2015.

Nibert, David. *Animal Oppression and Human Violence: Domesecration, Capitalism, and Global Conflict*. New York: Columbia University Press, 2013.

Parker, Ian. "Killing Animals at the Zoo." *The New Yorker*, January 16, 2017.

Pena-Guzman, David M. "Can non-human animals commit suicide?" *Animal Sentience*, 20, no. 1 (2017).

Regan, Tom. *Empty Cages: Facing the Challenge of Animal Rights*. London: Rowman and Littlefield, 2005.

Sanchez, Guadalupe, et al. "Human (Clovis)–gomphothere (*Cuvieronius* sp.) association 13,390 calibrated yBP in Sonora, Mexico." *PNAS* 111, no. 30 (July 29, 2014): 10972–77.

Singer, Peter. *Animal Liberation*. New York: Harper Collins, 1975.

Stuart, Diana, Rebecca L. Schewe, and Ryan Gunderson. "Extending Social Theory to Farm Animals: Addressing Alienation in the Dairy Sector." *Sociologia Ruralis* (2012): 1–22.

Wagoner, Brady, and Eric Jensen. "Science Learning at the Zoo: Evaluating Children's Developing Understanding of Animals and their Habitats." *Psychology and Society* 3, no. 1 (2010): 65–76.

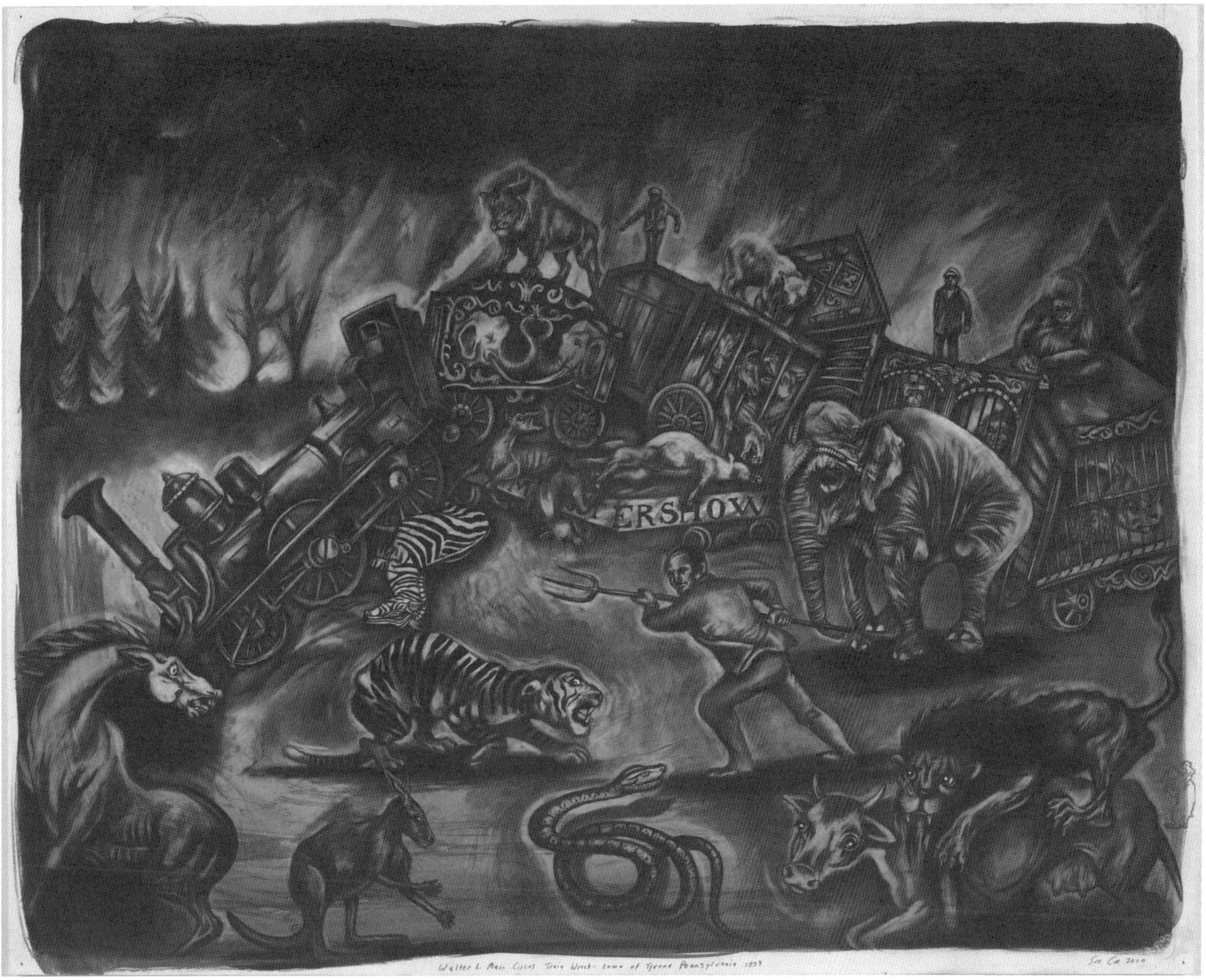
ERS.IOW
Walter L. Main Circus Train Wreck - town of Tyrone Pennsylvania 1893
Sue Coe 2010

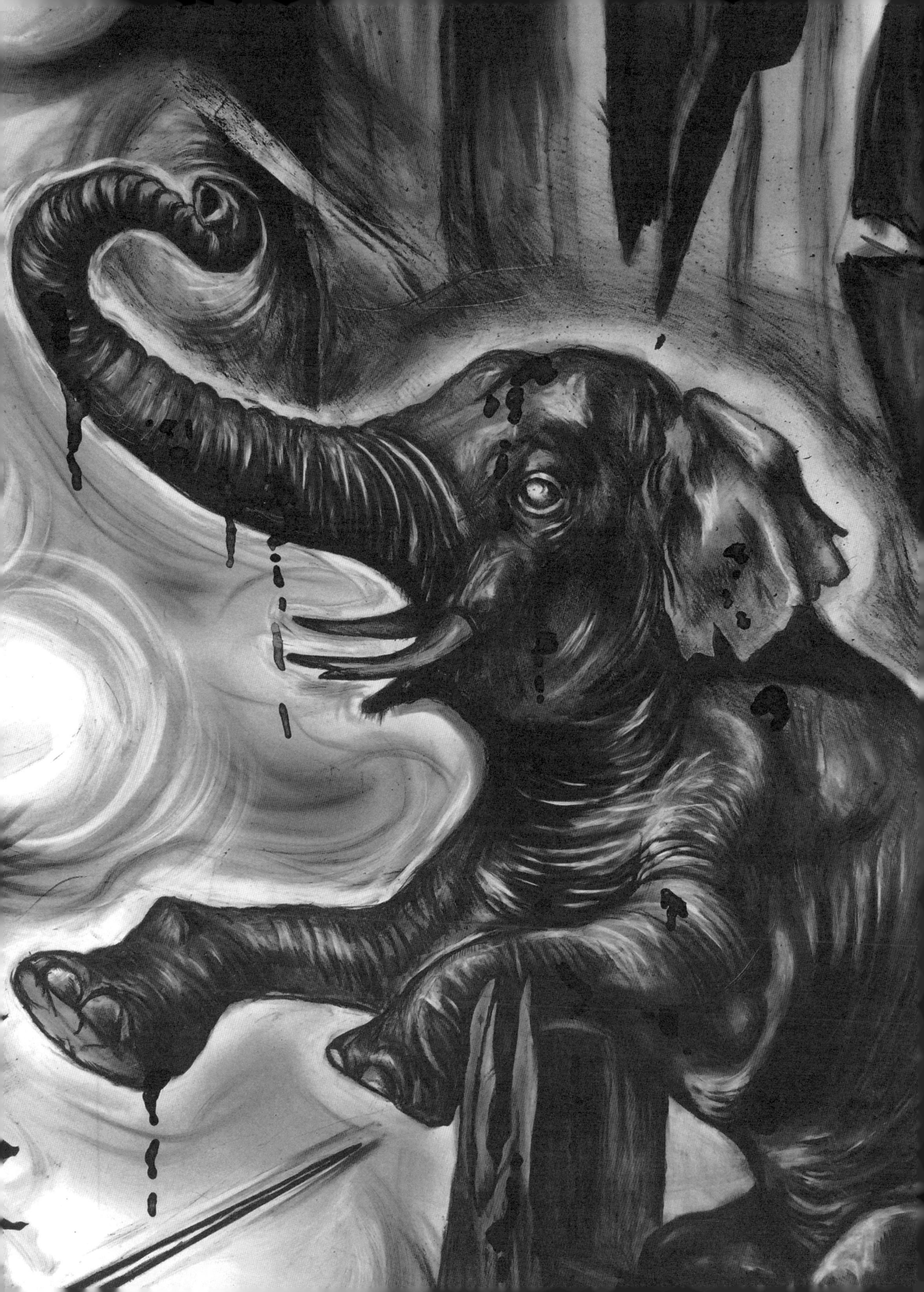

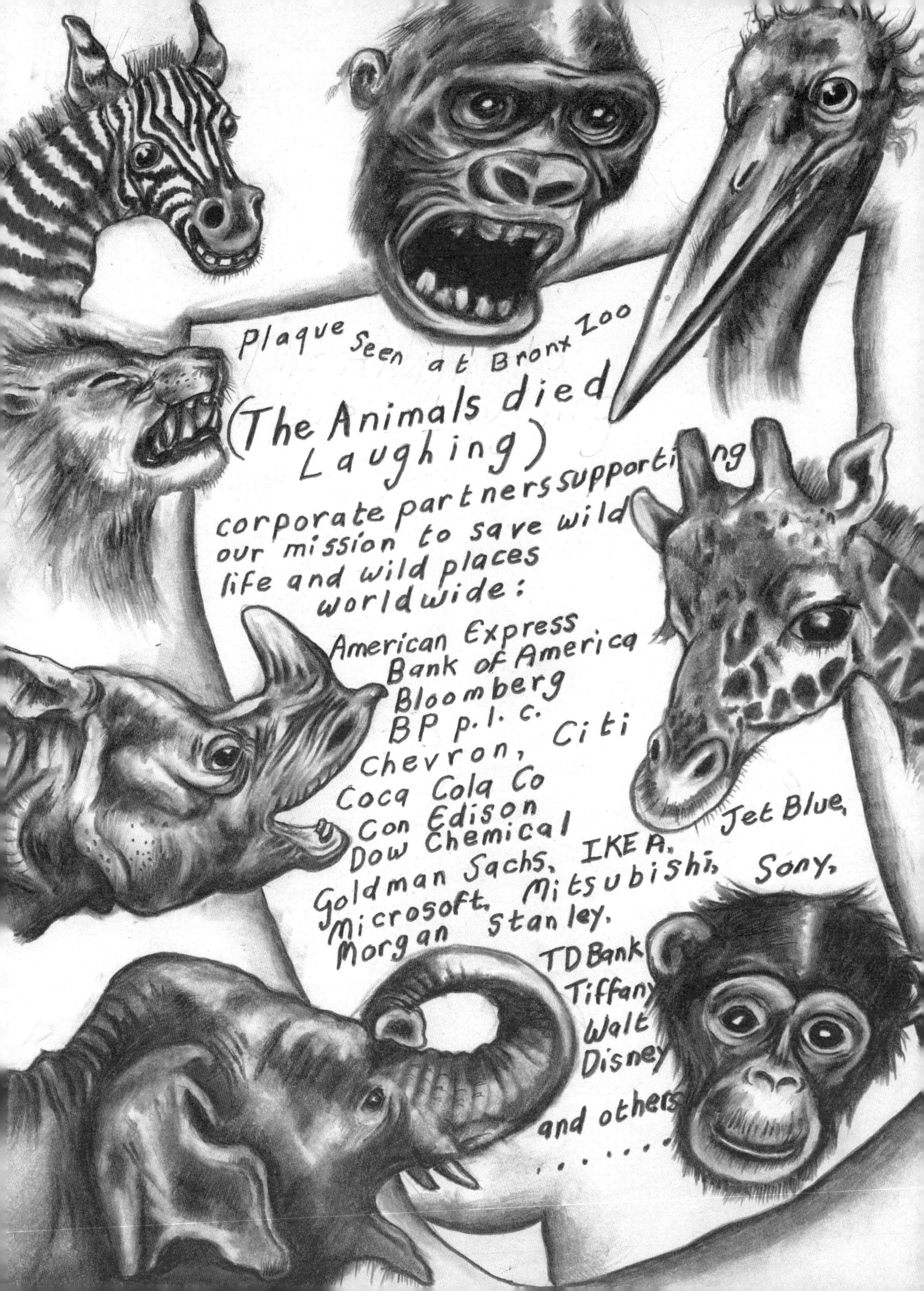
Plaque Seen at Bronx Zoo
(The Animals died Laughing)
corporate partners supporting our mission to save wild life and wild places worldwide:
American Express
Bank of America
Bloomberg
BP p.l.c.
Citi
Chevron,
Coca Cola Co
Con Edison
Dow Chemical
Jet Blue,
Goldman Sachs,
IKEA,
Mitsubishi,
Sony,
Microsoft,
Morgan Stanley,
TD Bank
Tiffany
Walt Disney
and others......

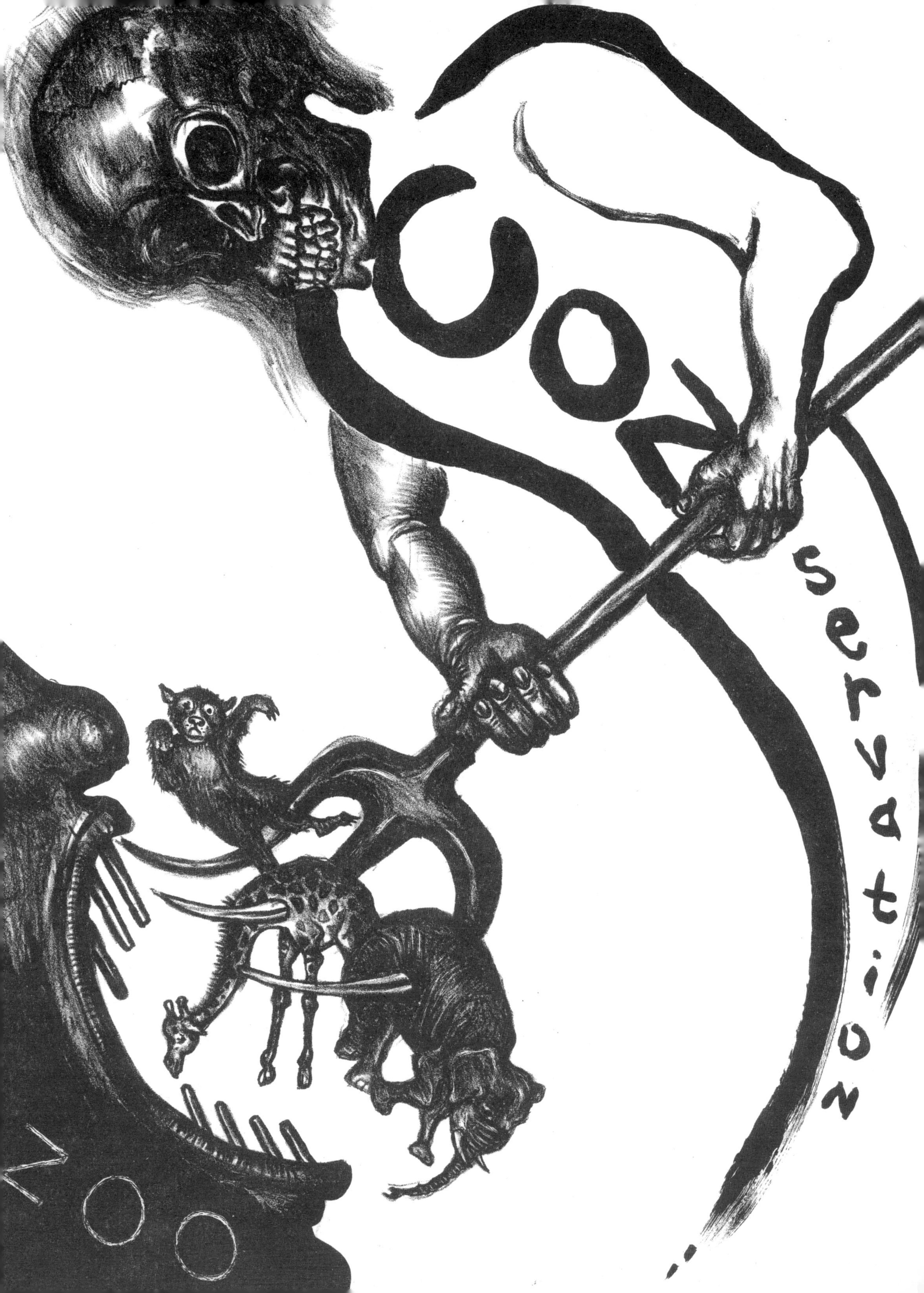
Con
servation
ZOO

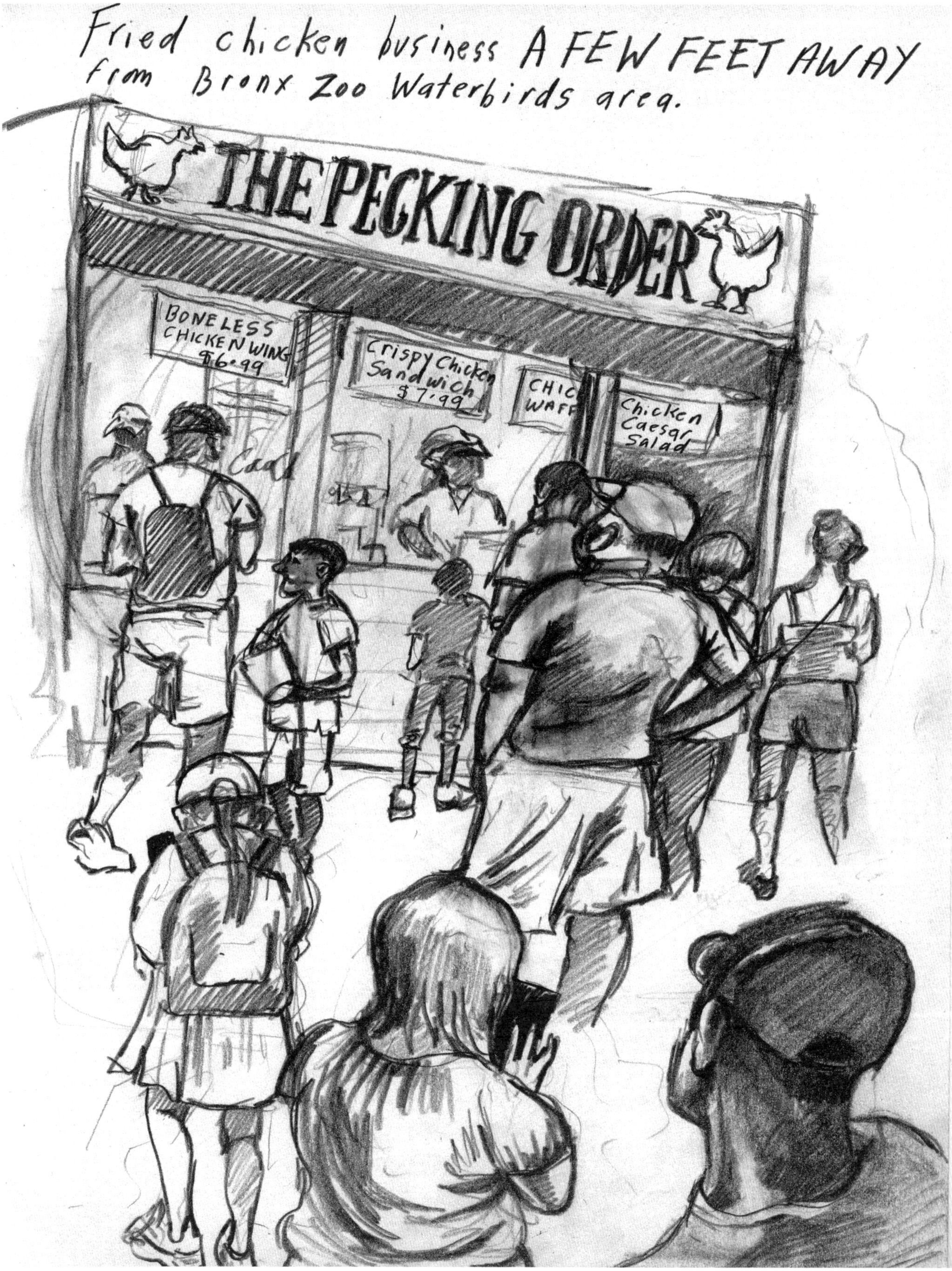
Fried chicken business A FEW FEET AWAY from Bronx Zoo Waterbirds area.
THE PECKING ORDER
BONELESS CHICKEN WINGS $6.99
Crispy Chicken Sandwich $7.99
Chicken Caesar Salad

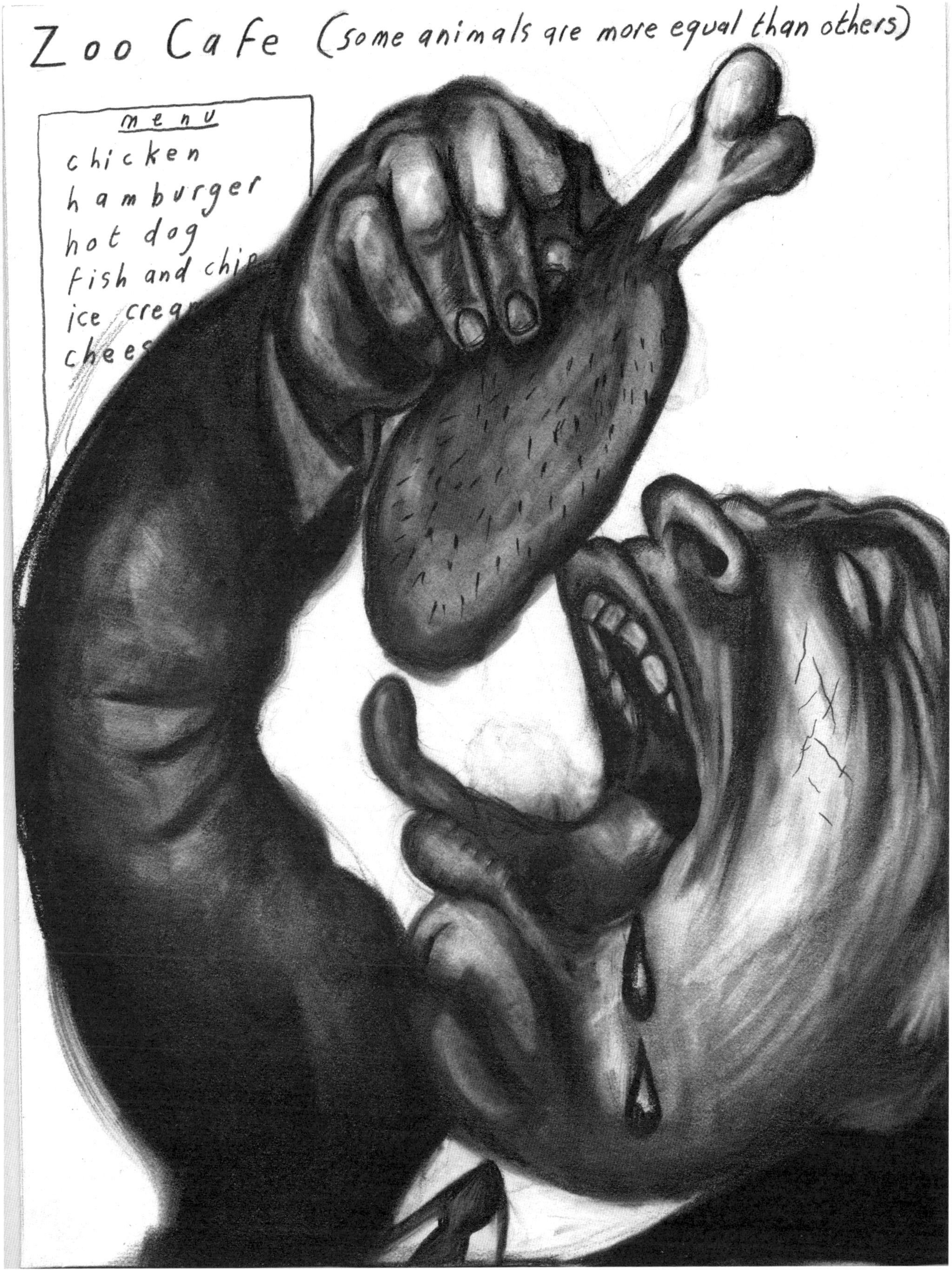
Zoo Cafe (some animals are more equal than others)
menu
chicken
hamburger
hot dog

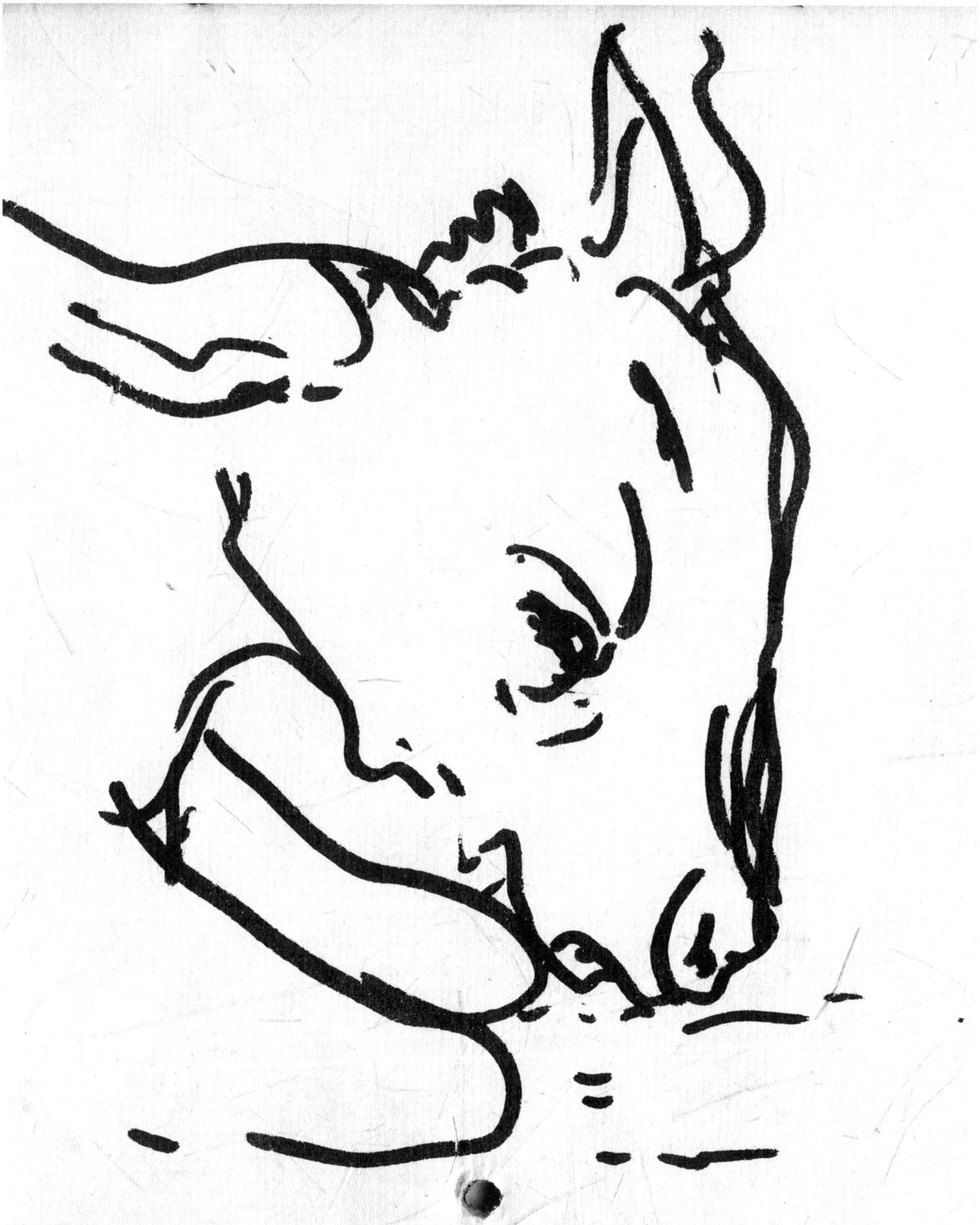

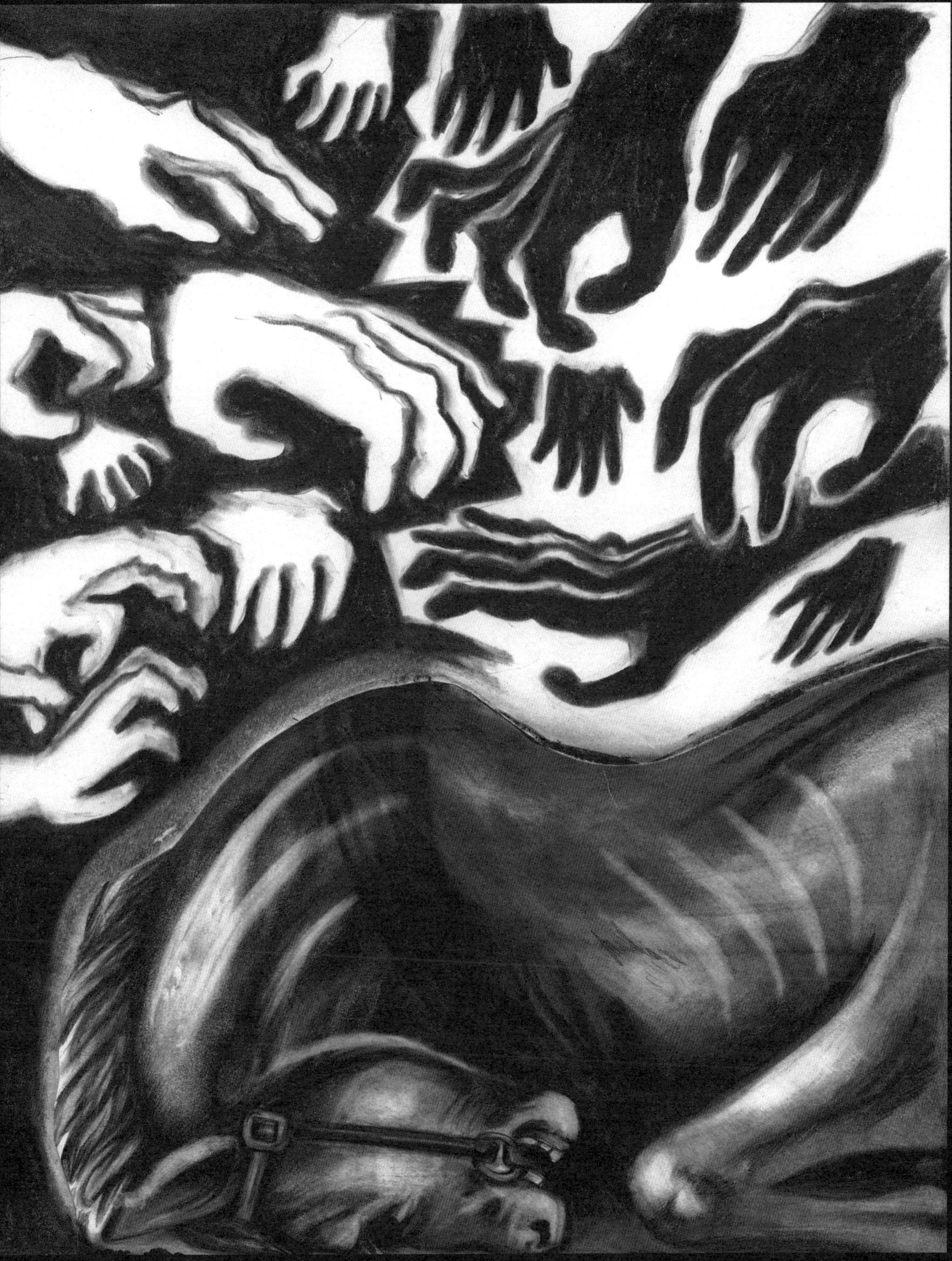

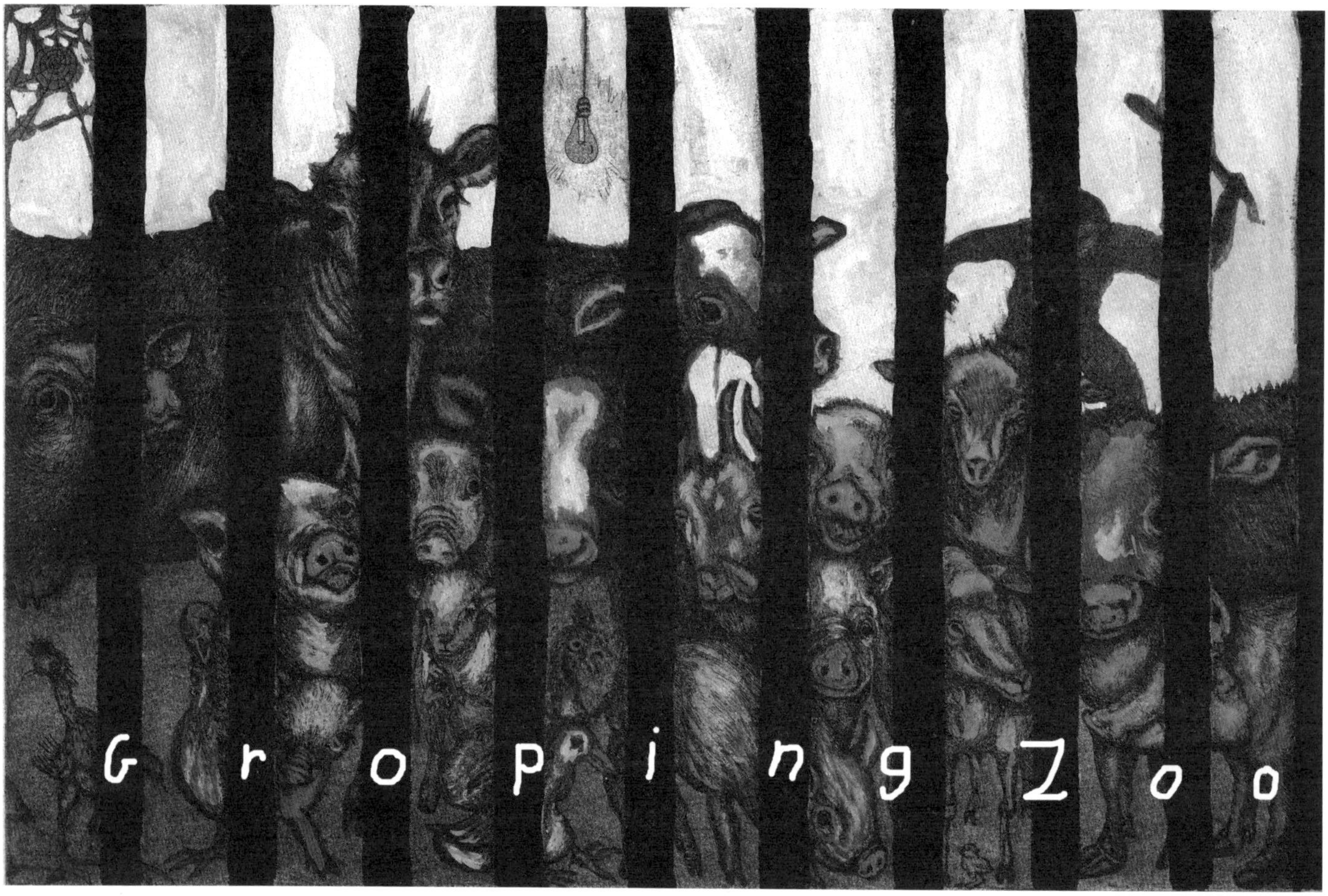
Groping Zoo

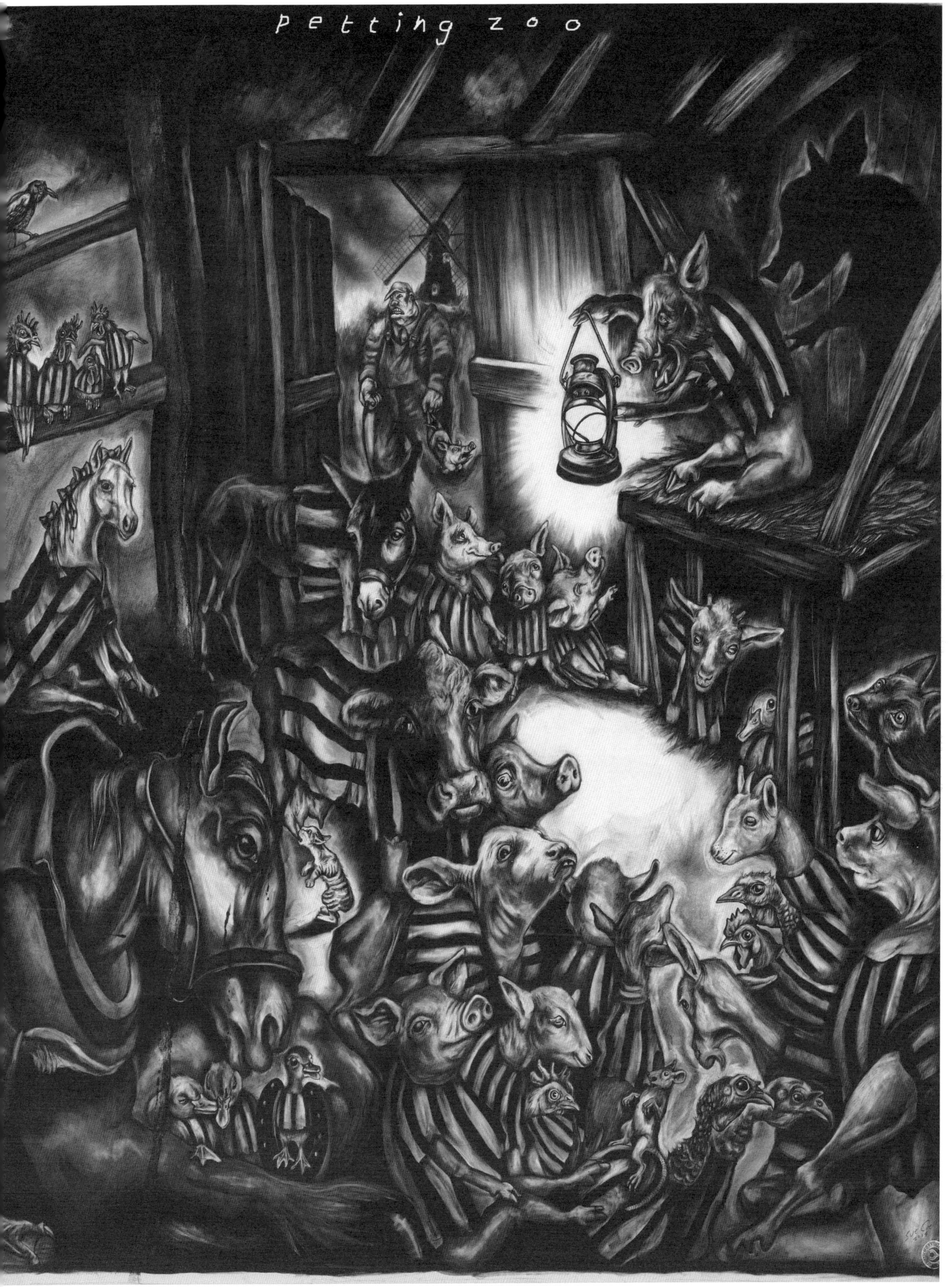
petting zoo

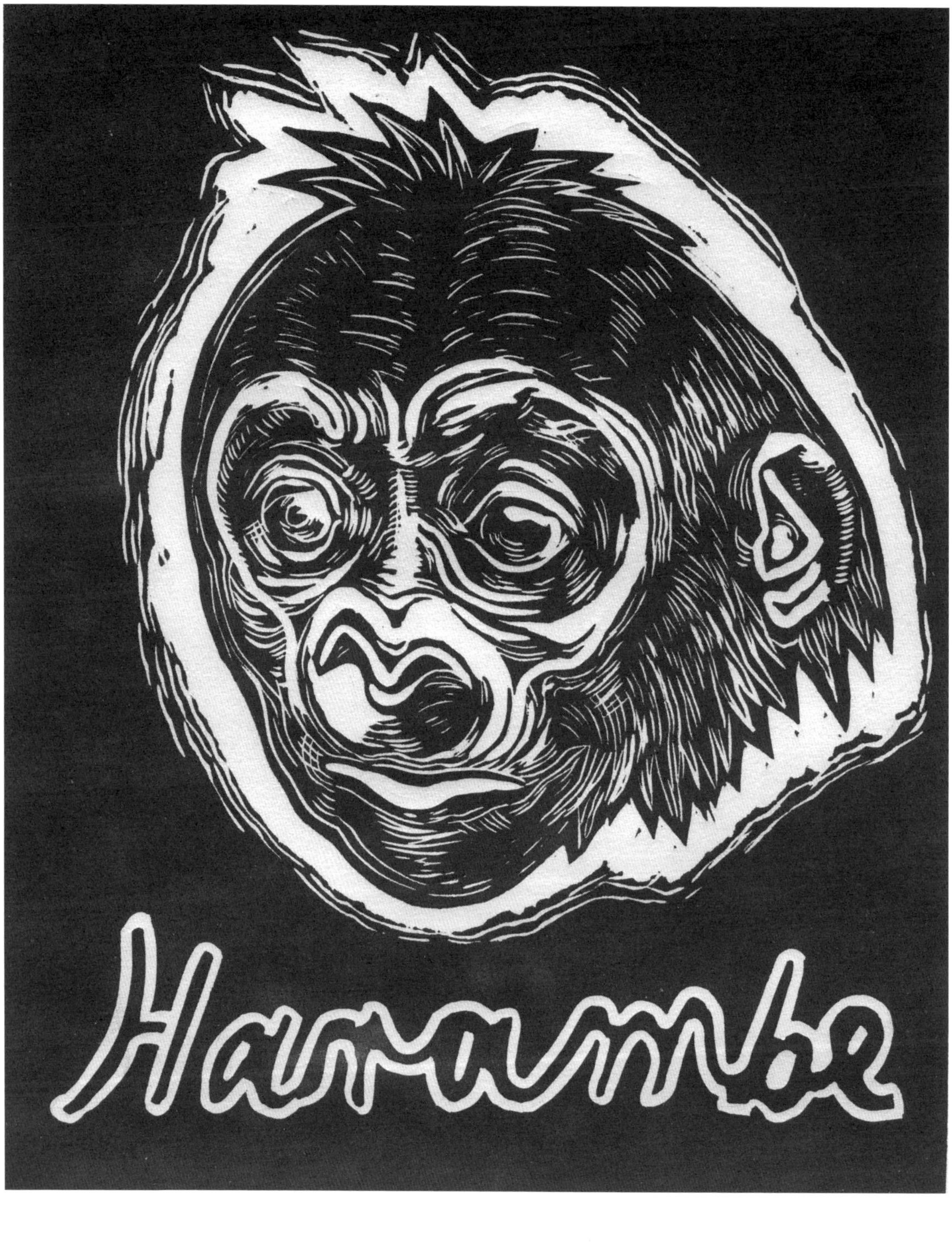
Harambe

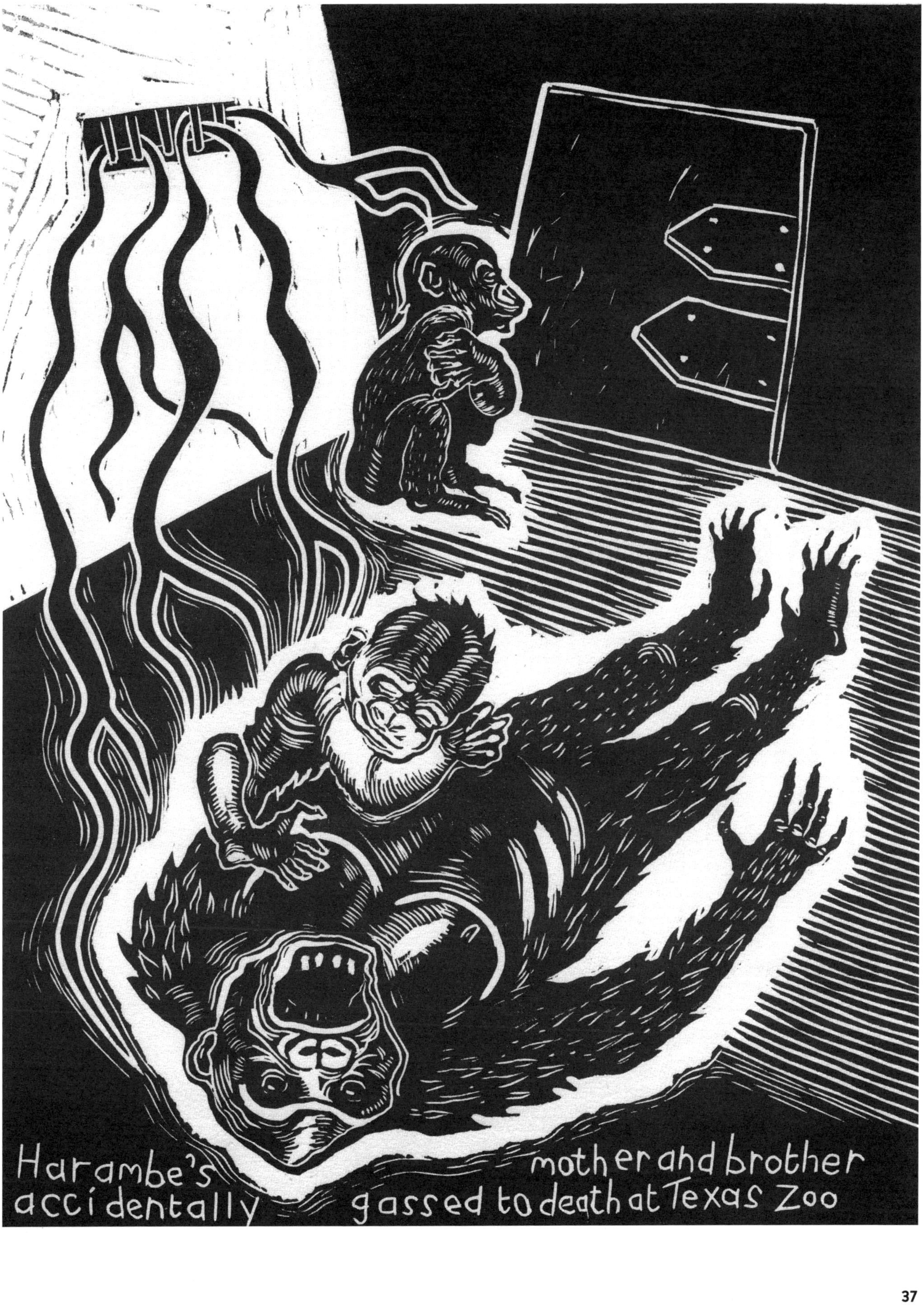
Harambe's
mother and brother
accidentally
gassed to death at Texas Zoo

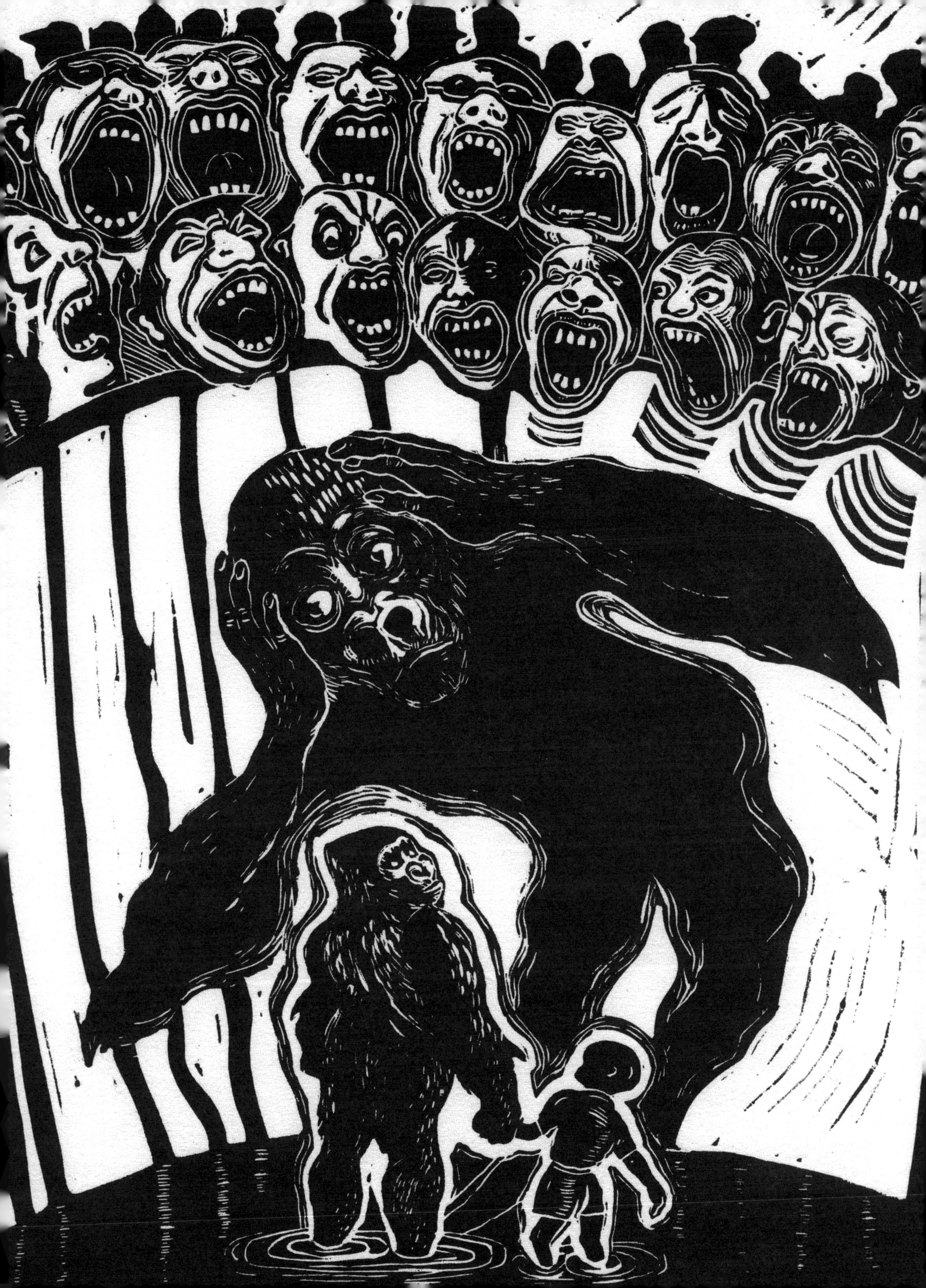

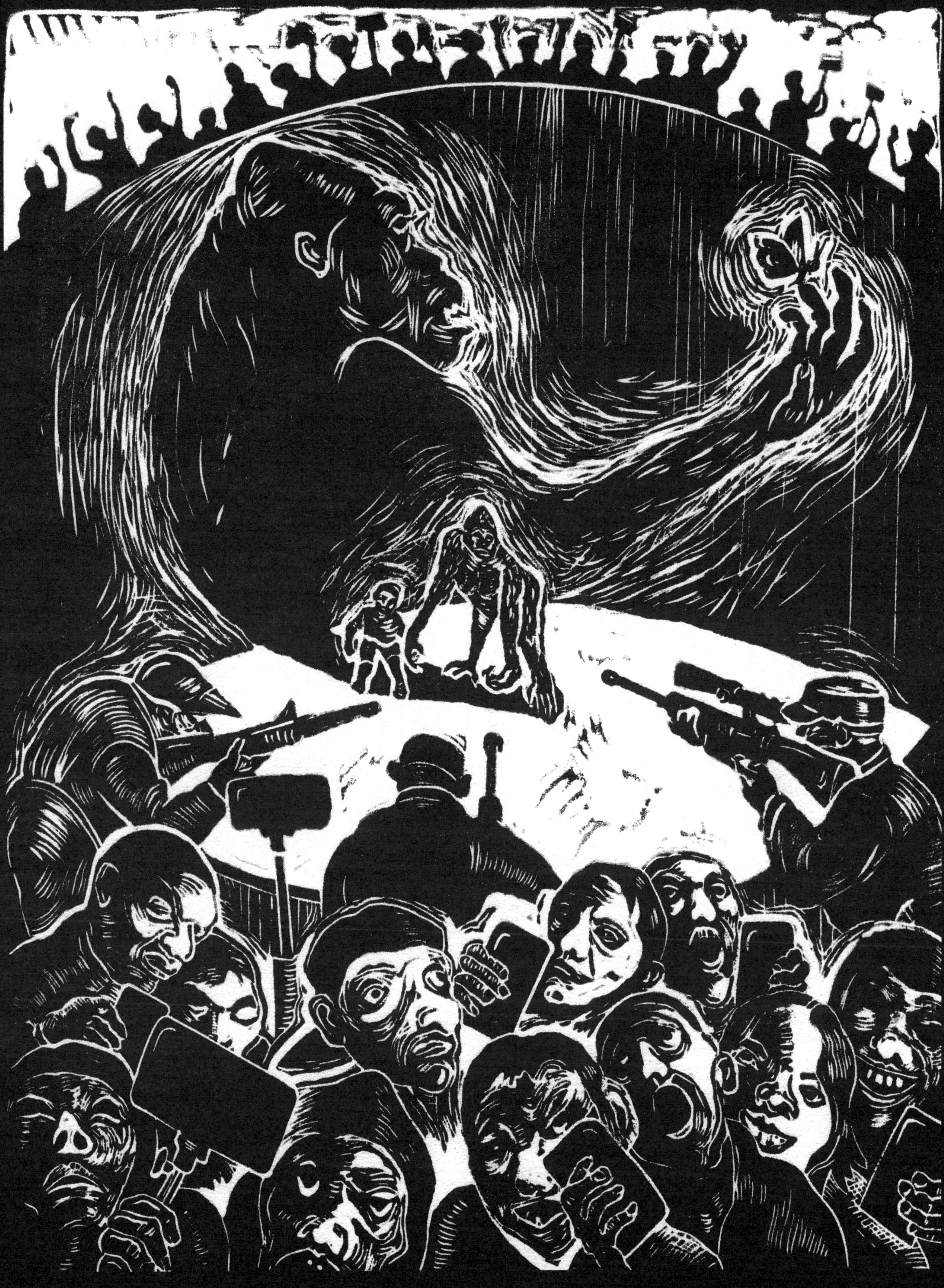

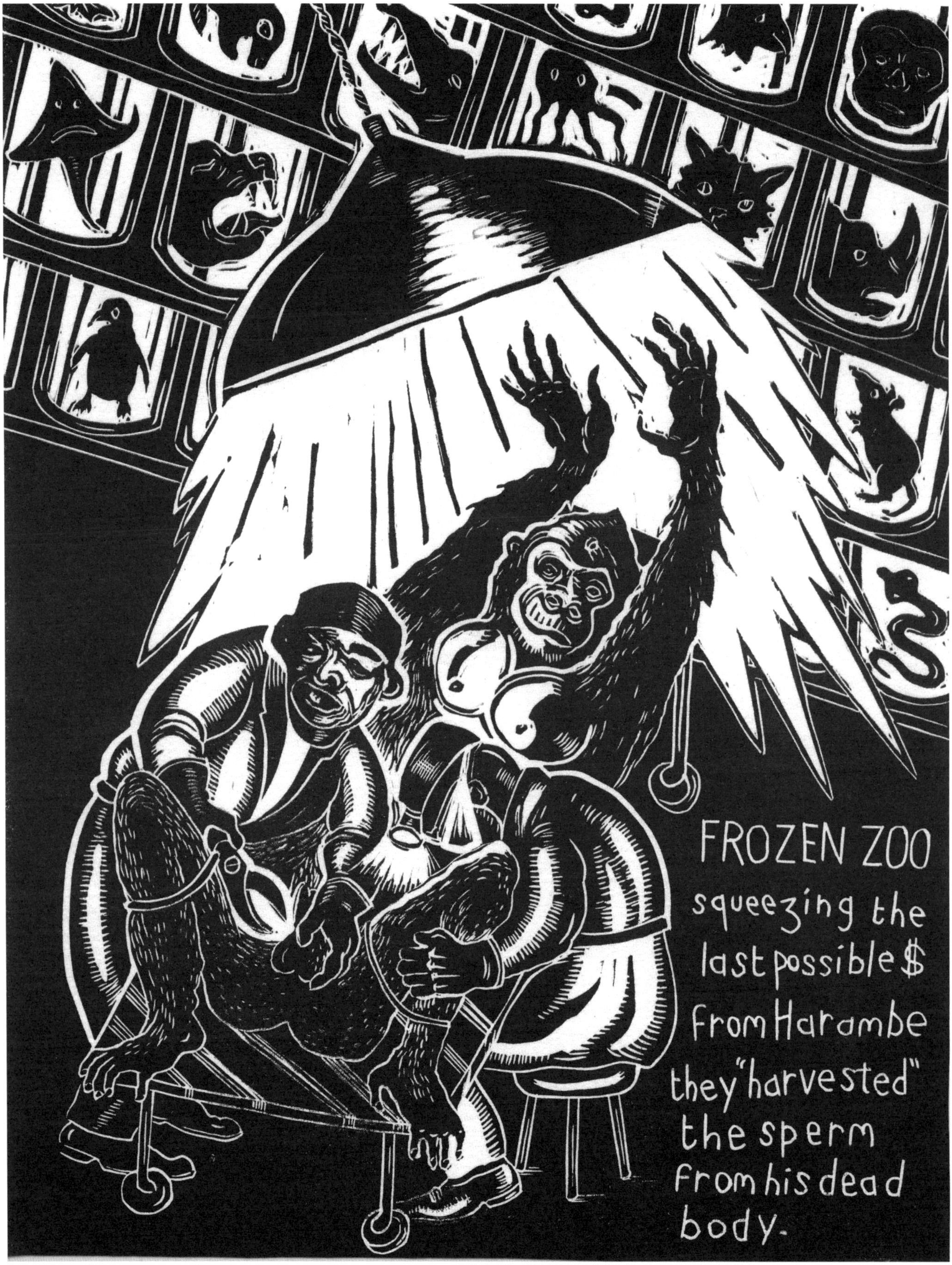
FROZEN ZOO
squeezing the last possible $ from Harambe they "harvested" the sperm from his dead body.

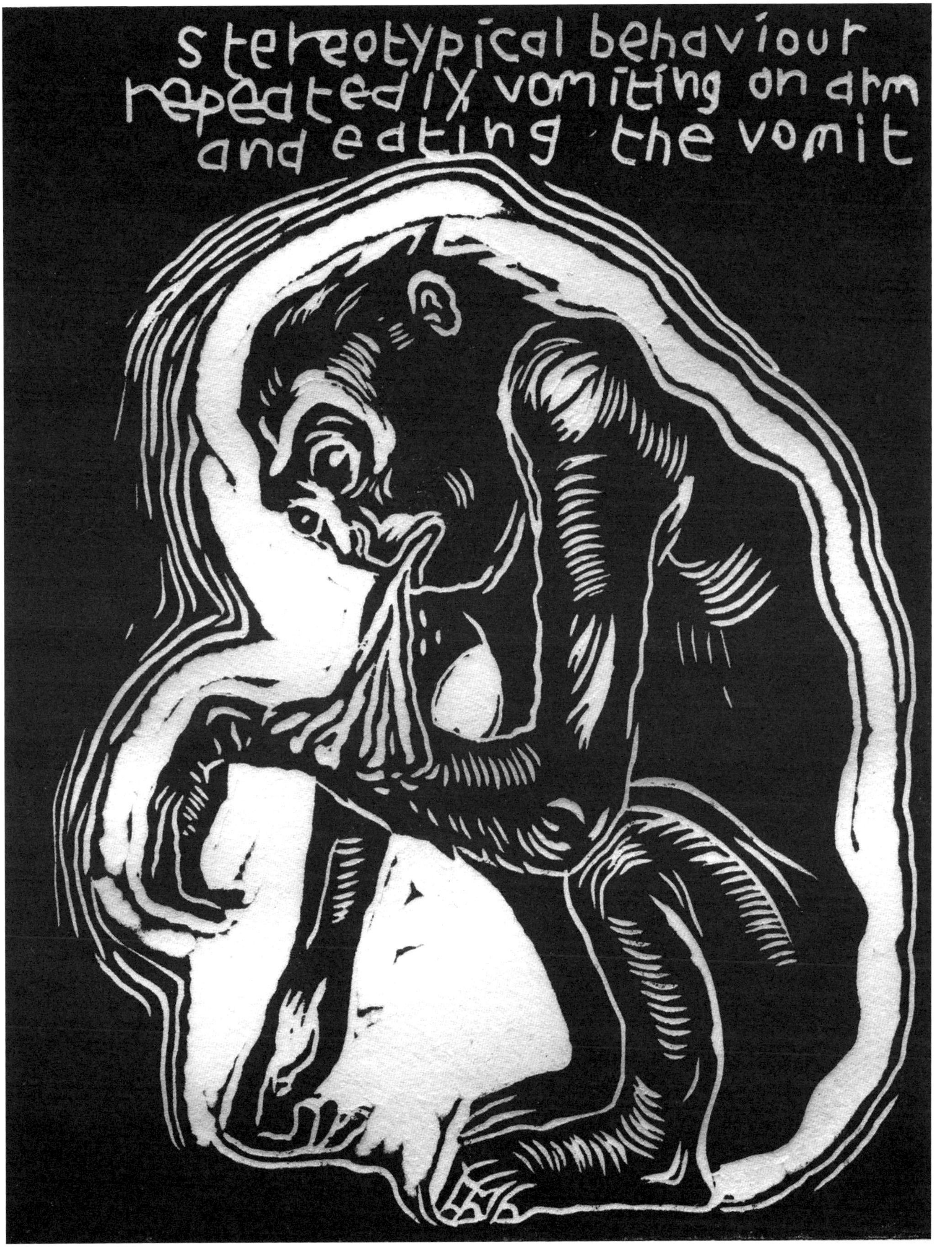
stereotypical behaviour
repeatedly vomiting on arm
and eating the vomit

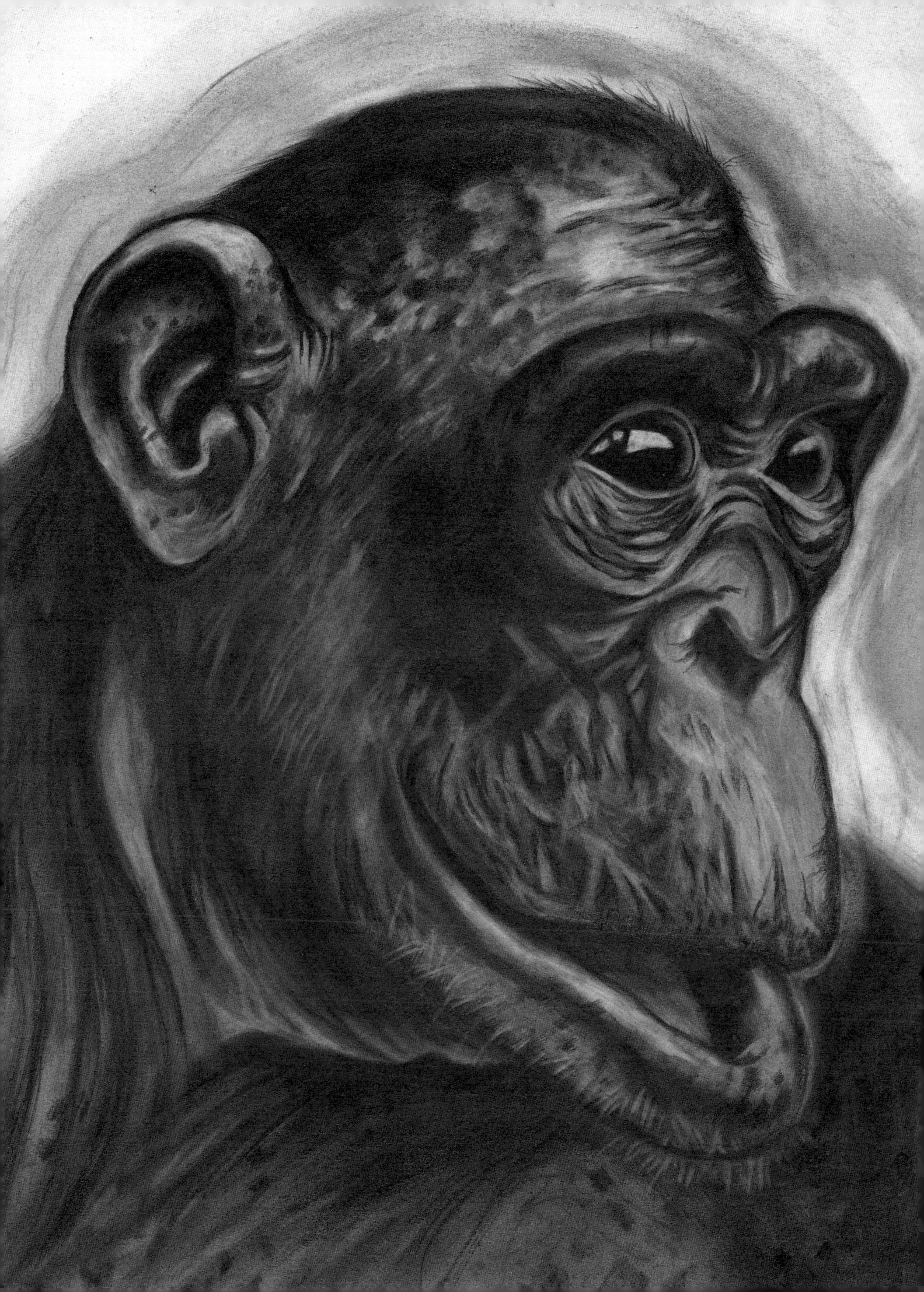

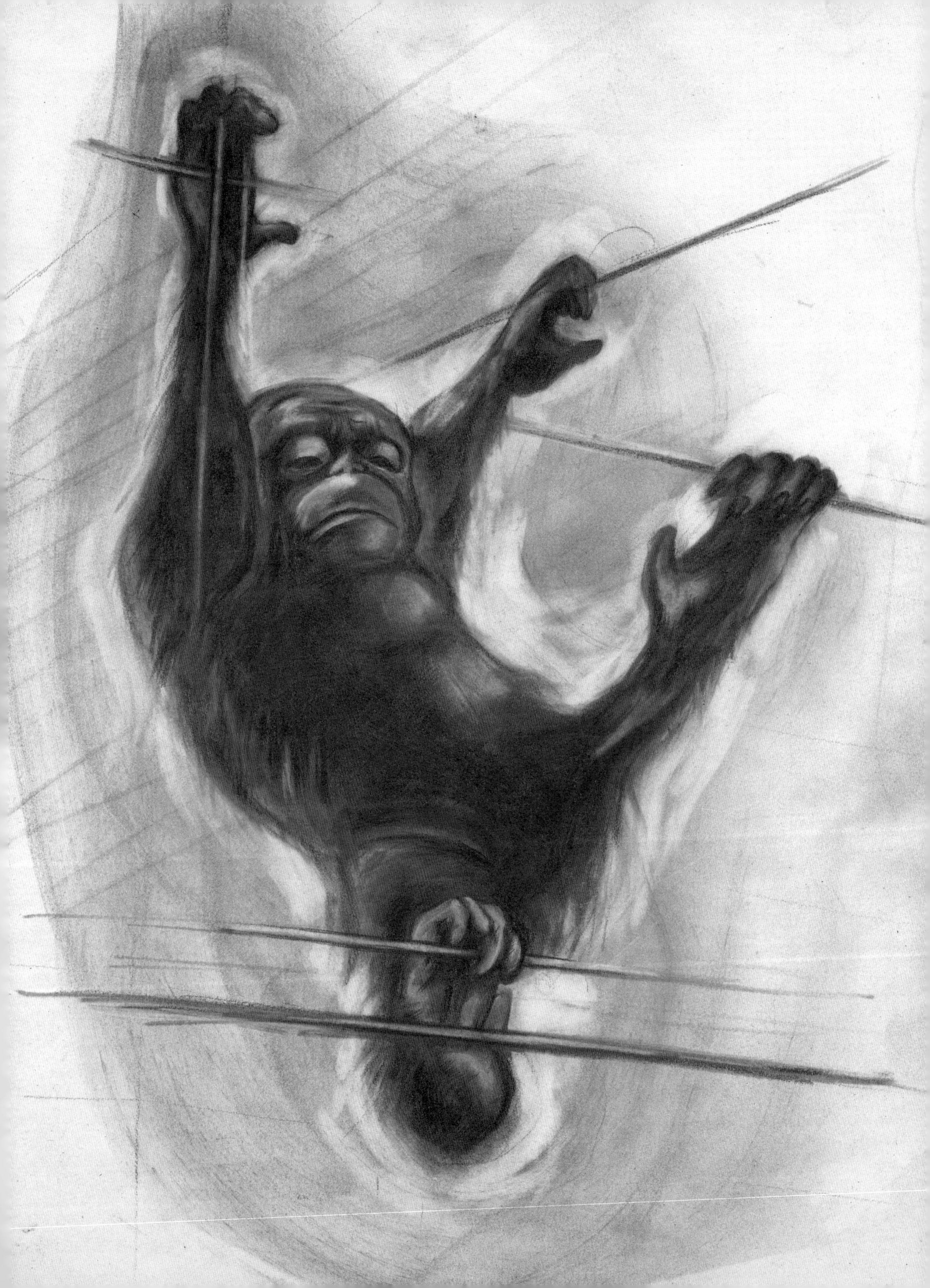

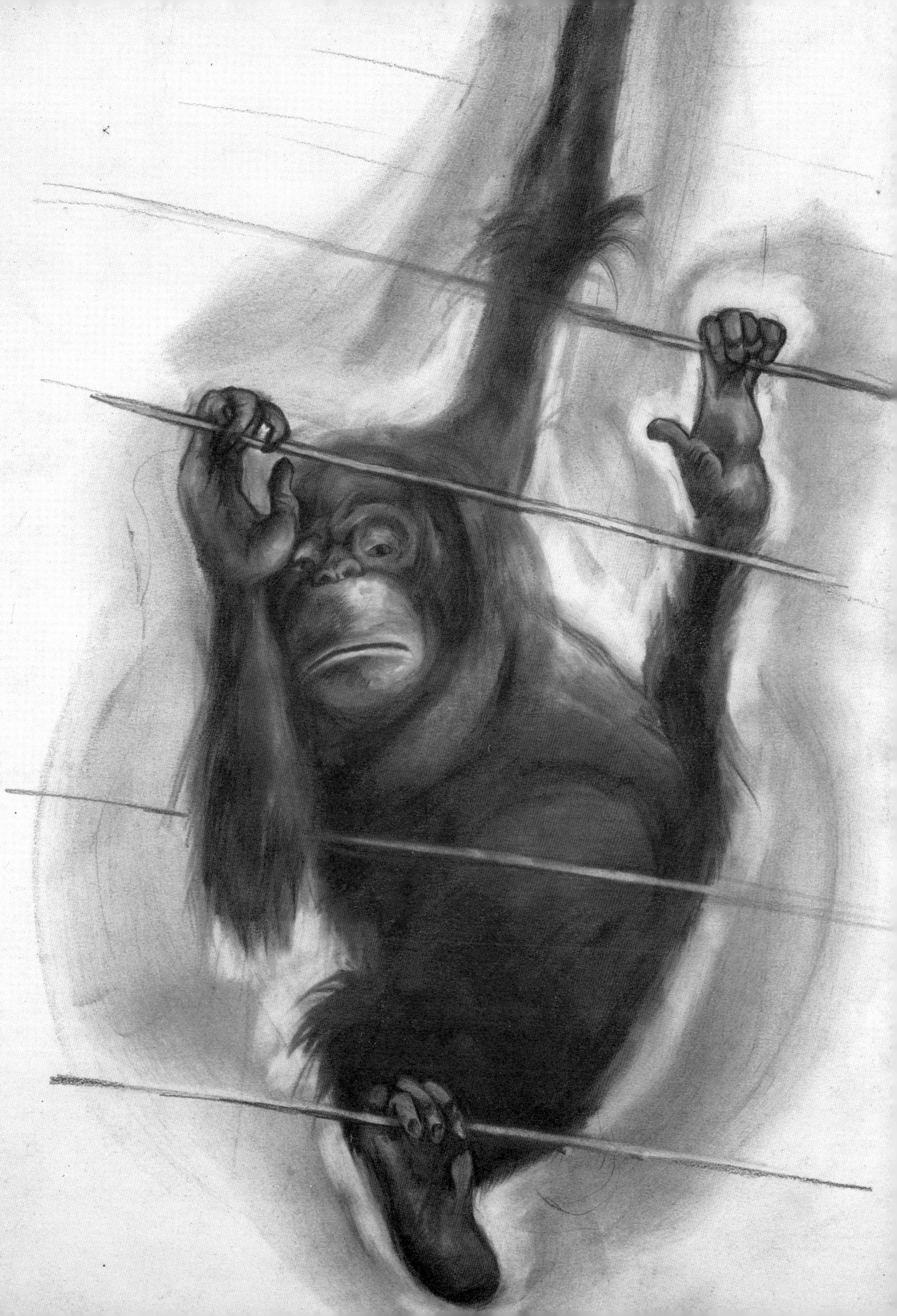

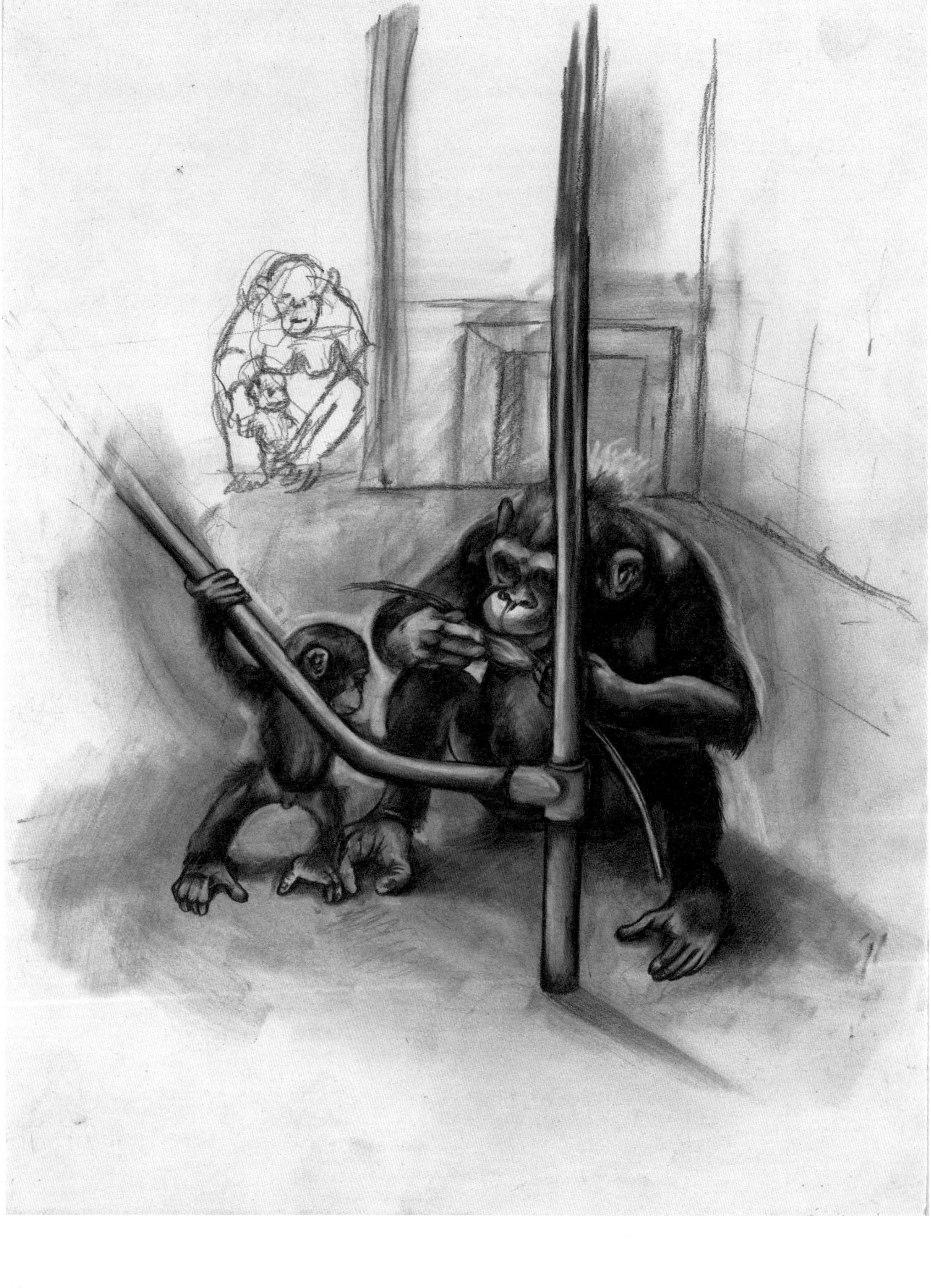

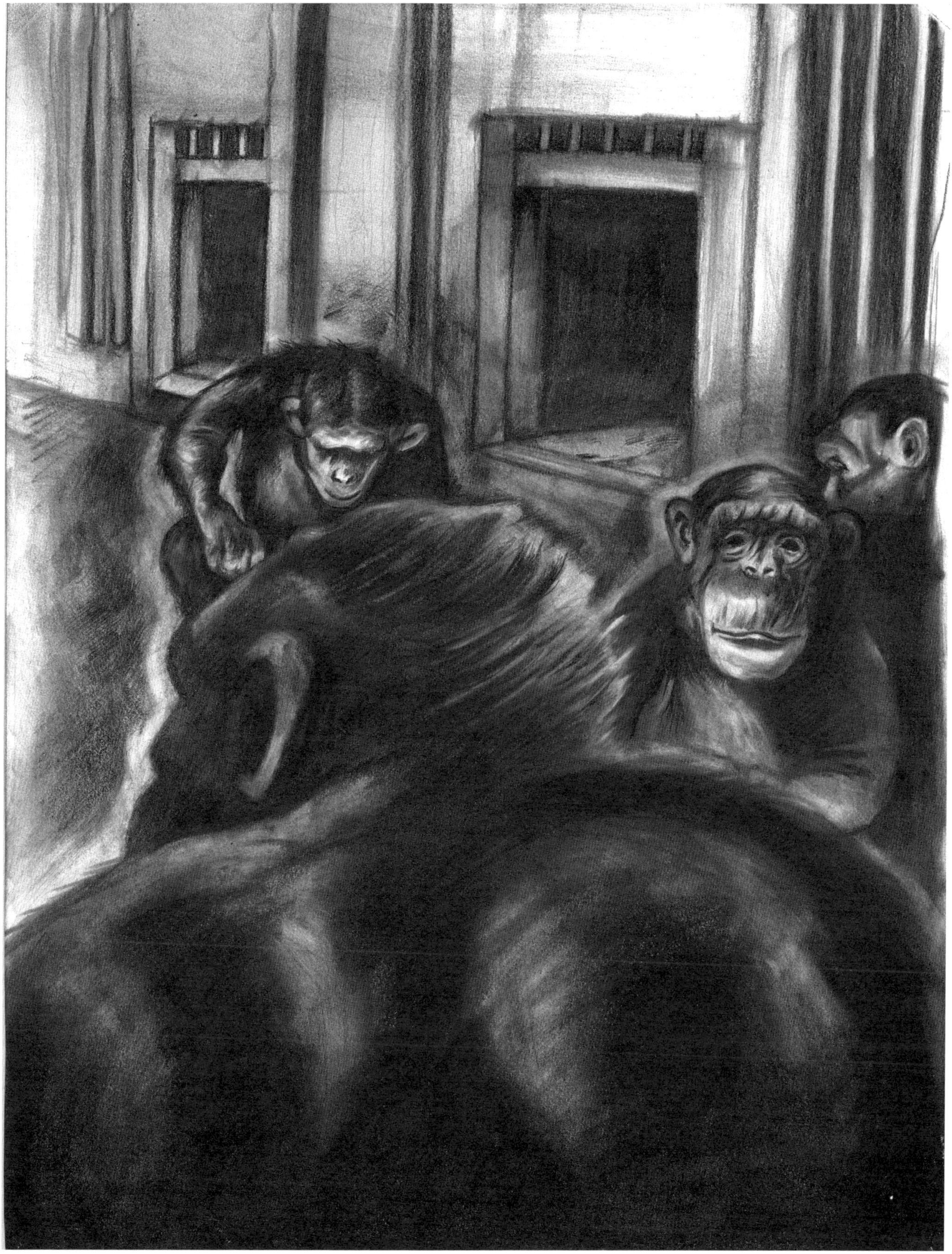

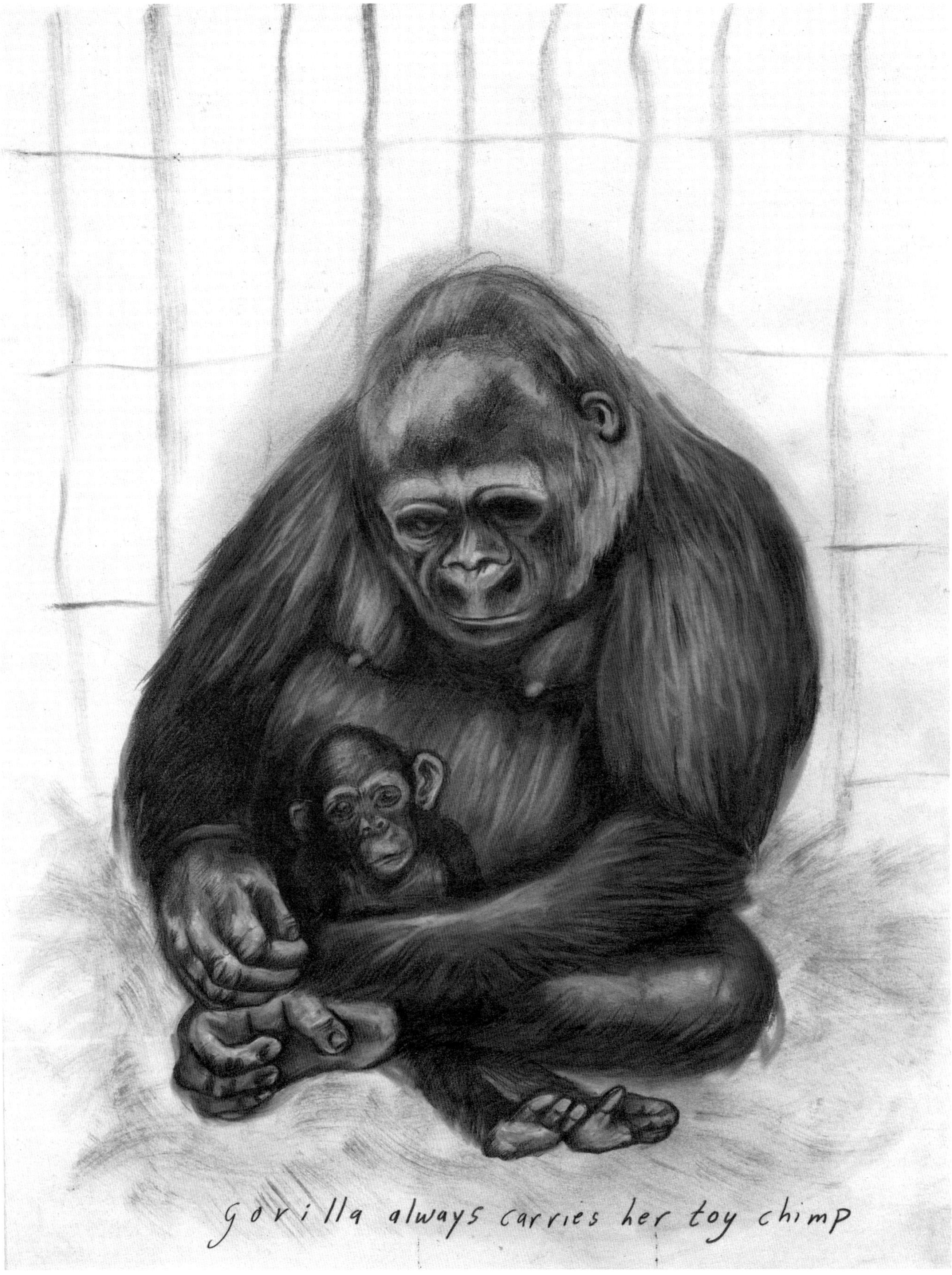
gorilla always carries her toy chimp

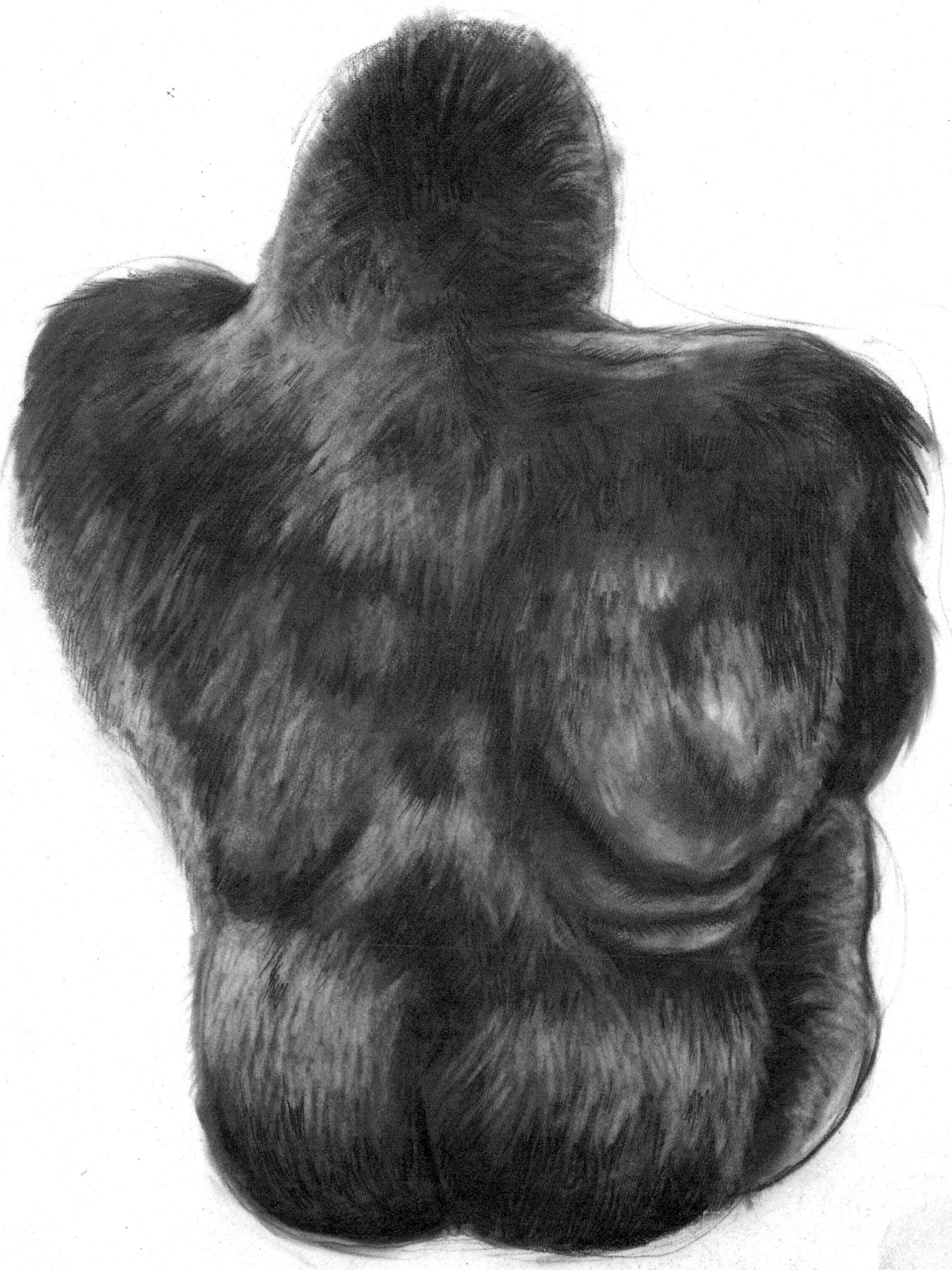

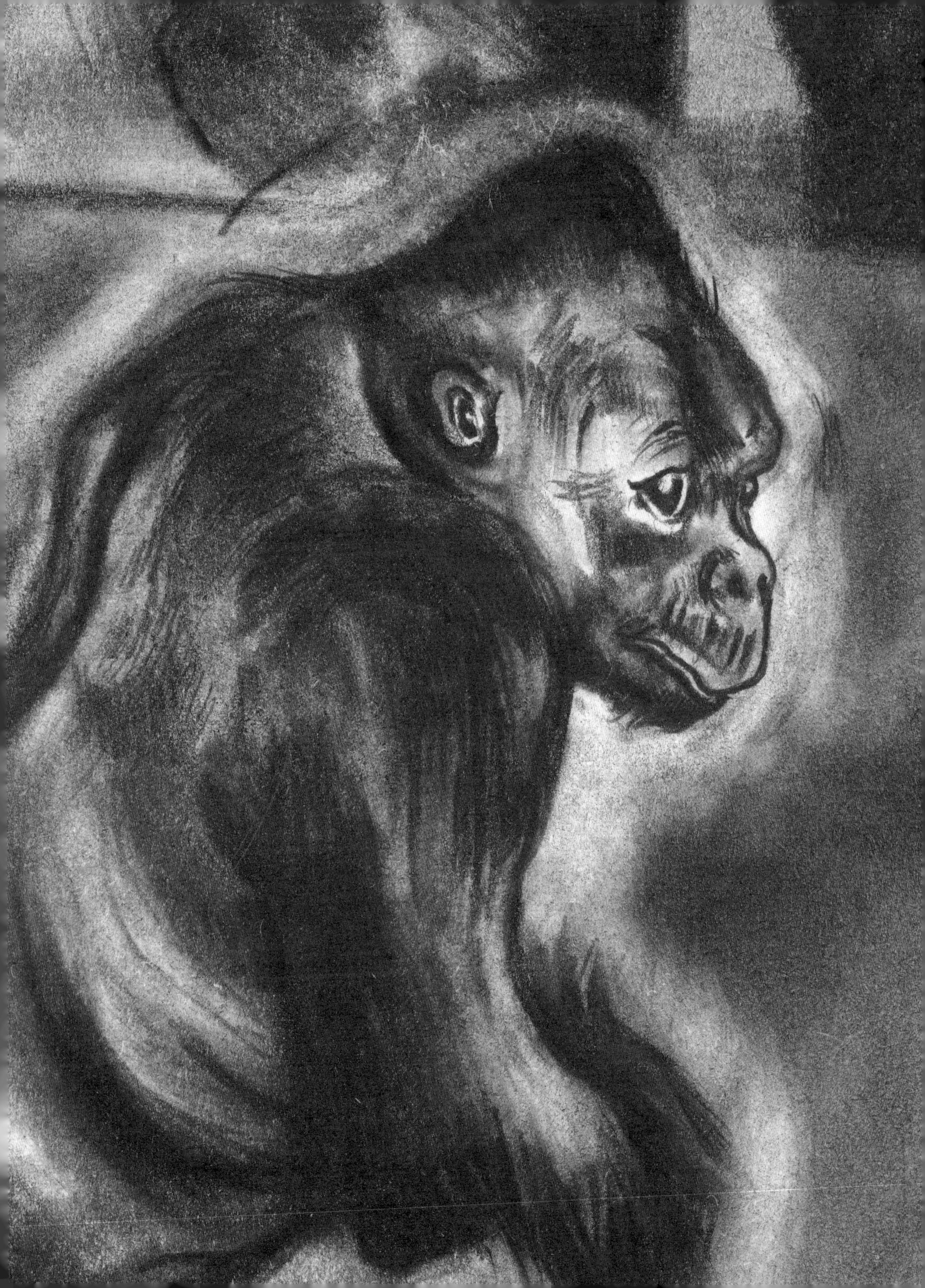

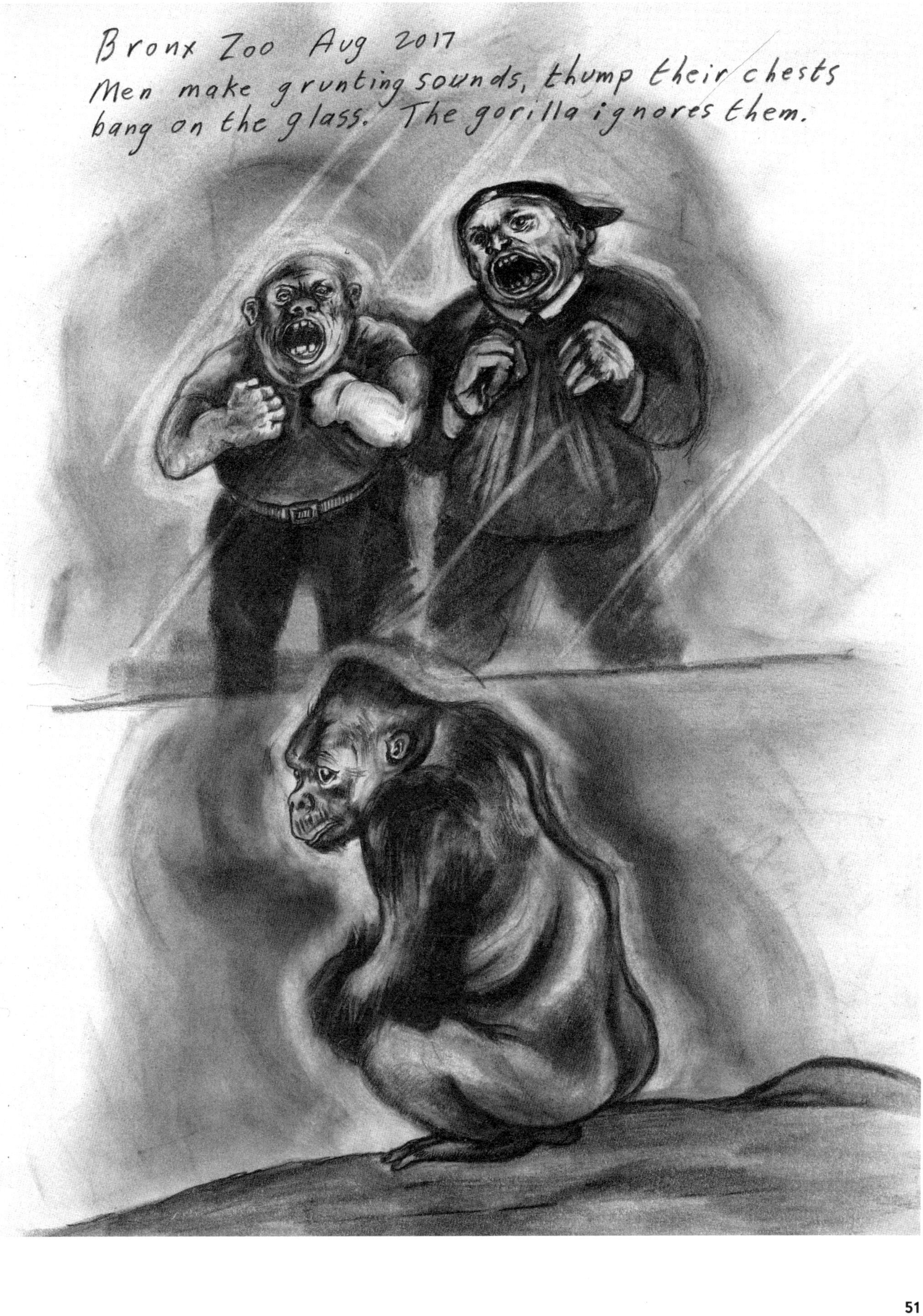
Bronx Zoo Aug 2017
Men make grunting sounds, thump their chests bang on the glass. The gorilla ignores them.

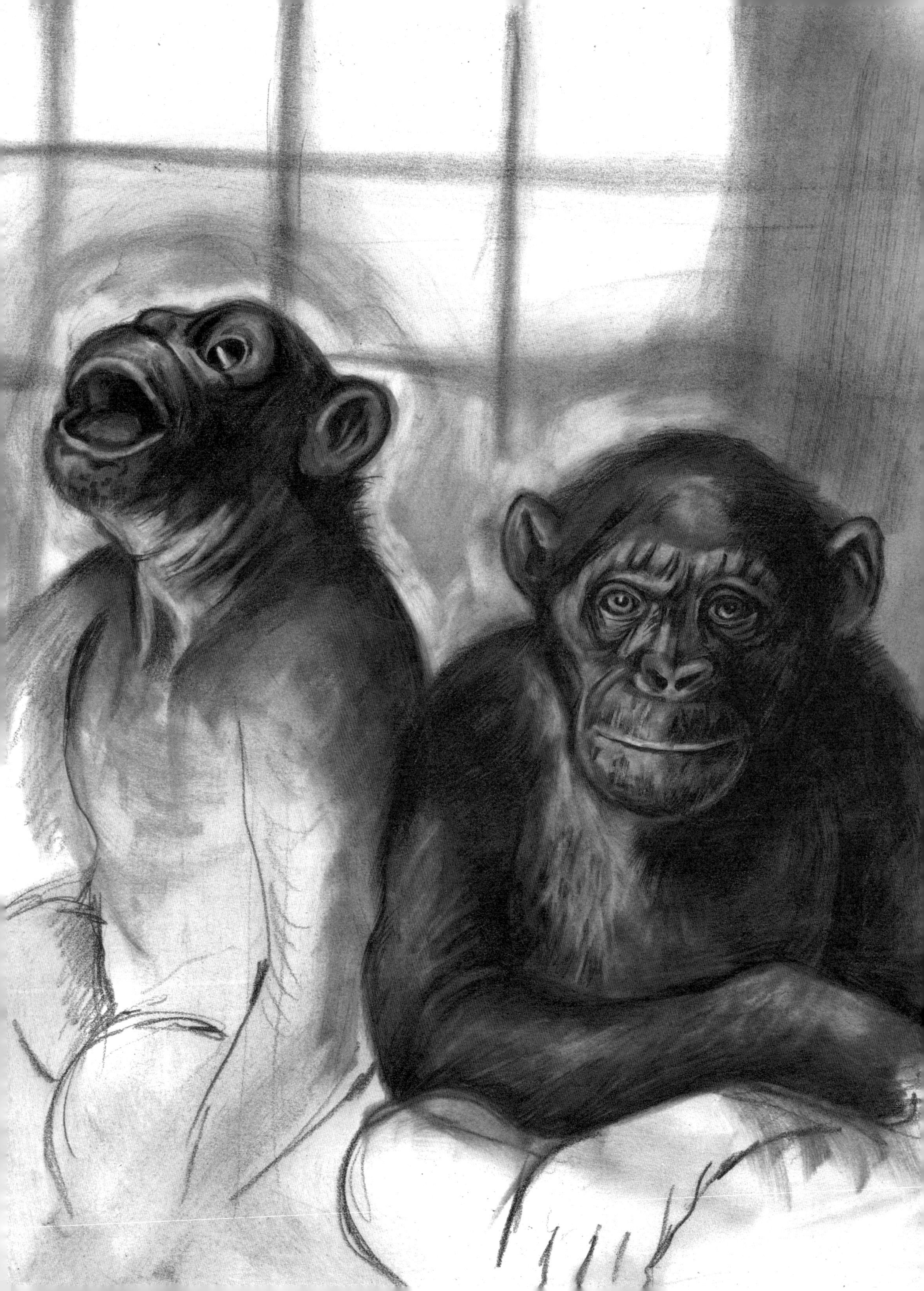

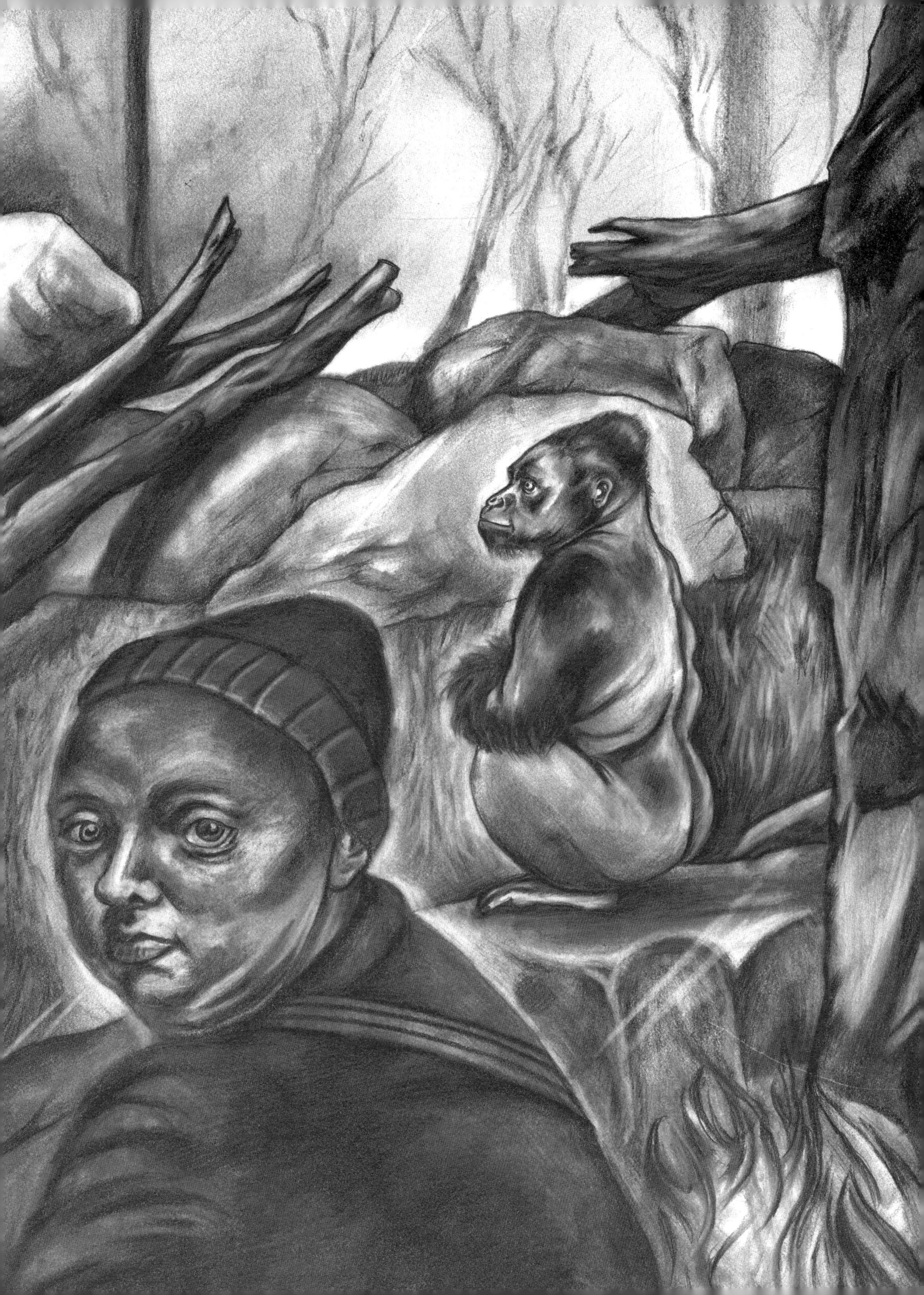

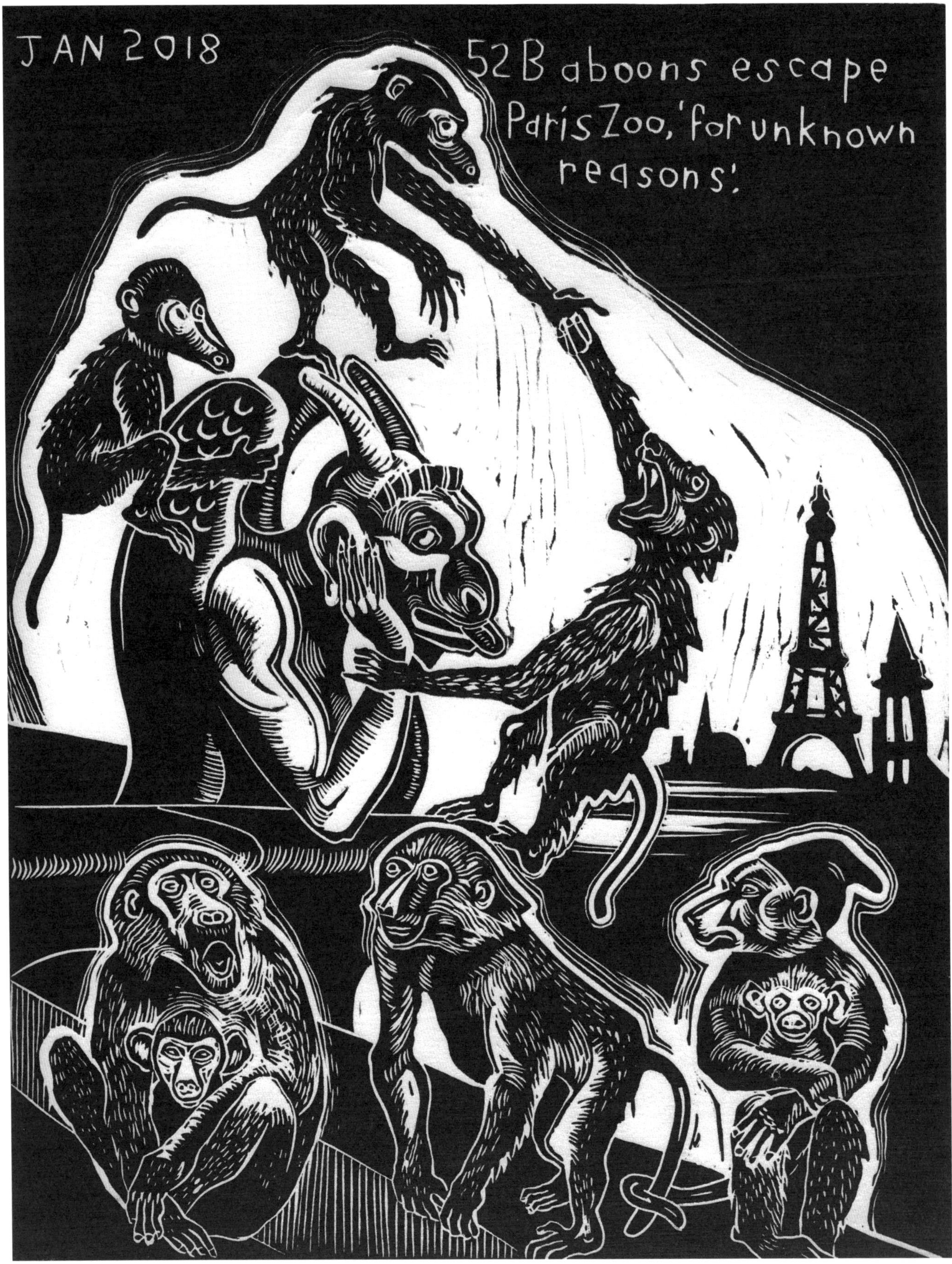
JAN 2018
52 Baboons escape
Paris Zoo, 'for unknown
reasons'.

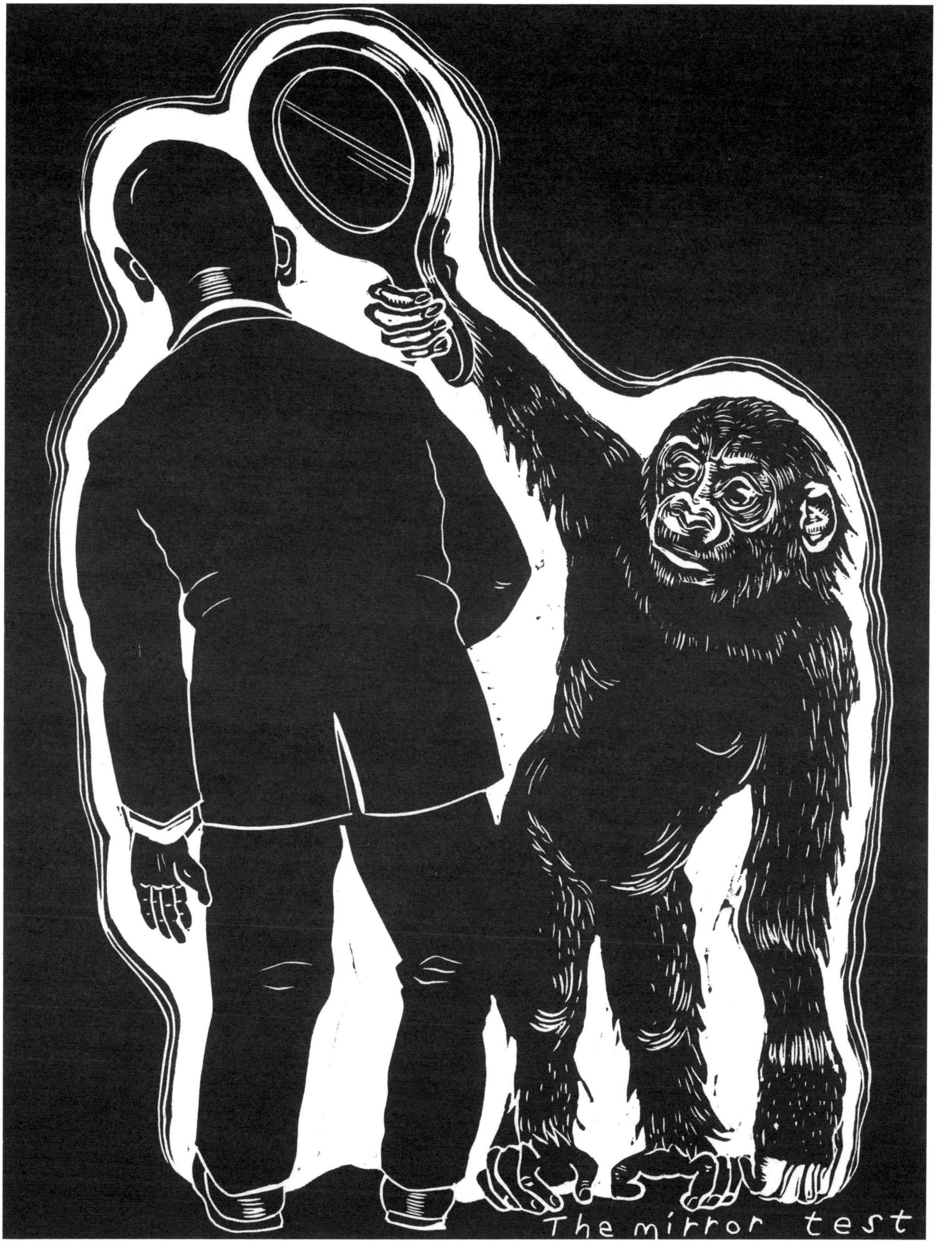
The mirror test

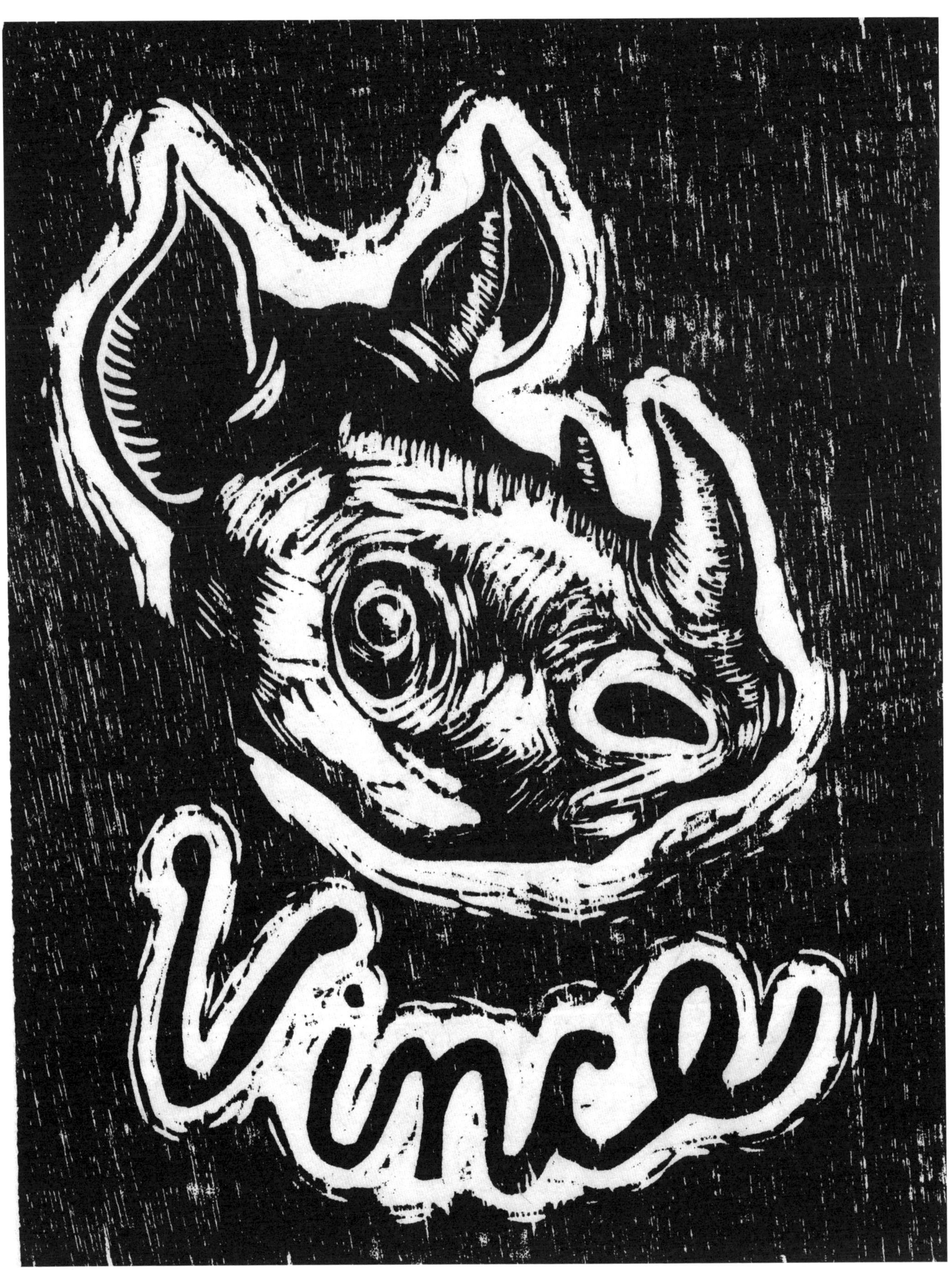
Vince

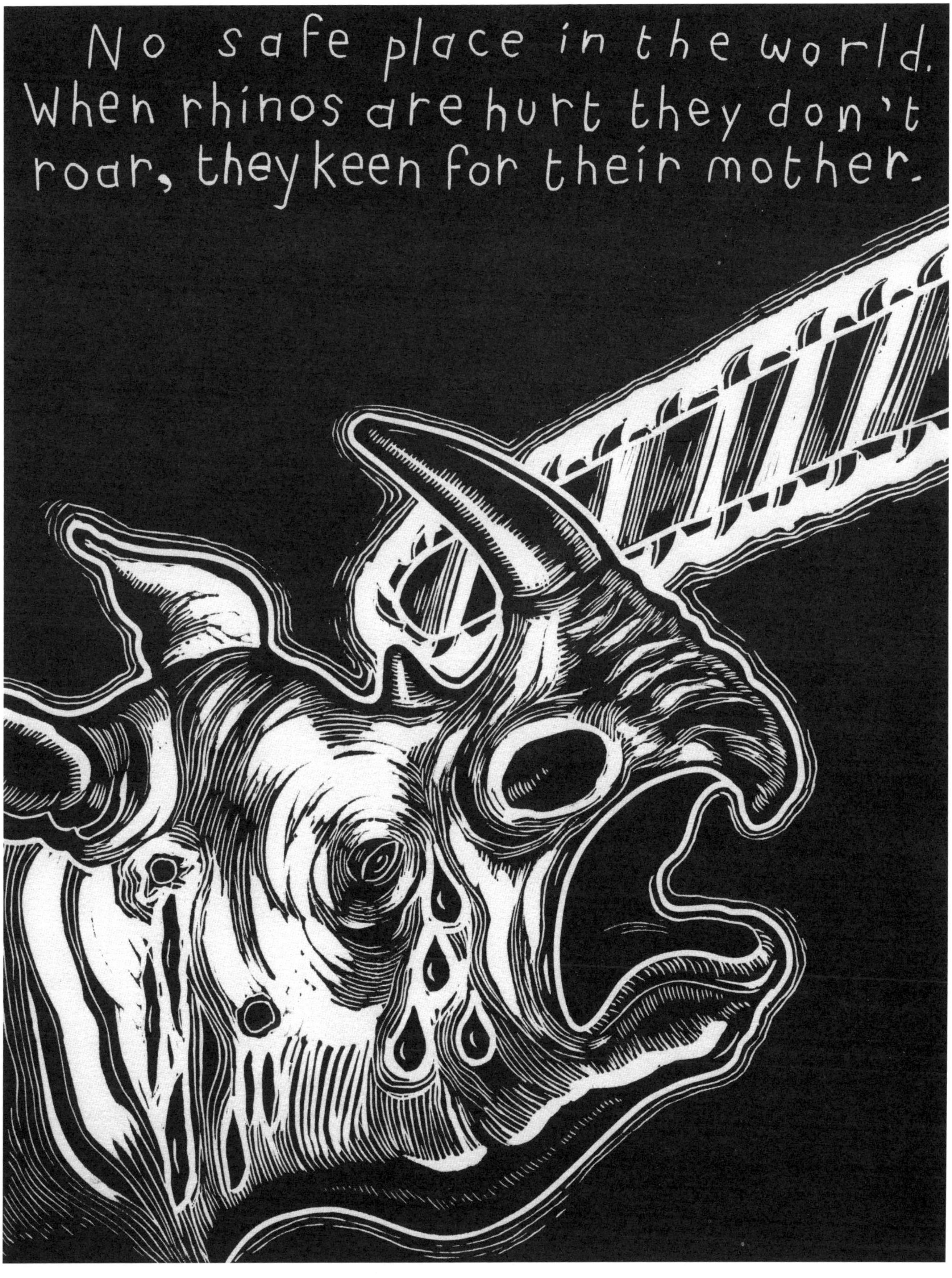
No safe place in the world.
When rhinos are hurt they don't roar, they keen for their mother.

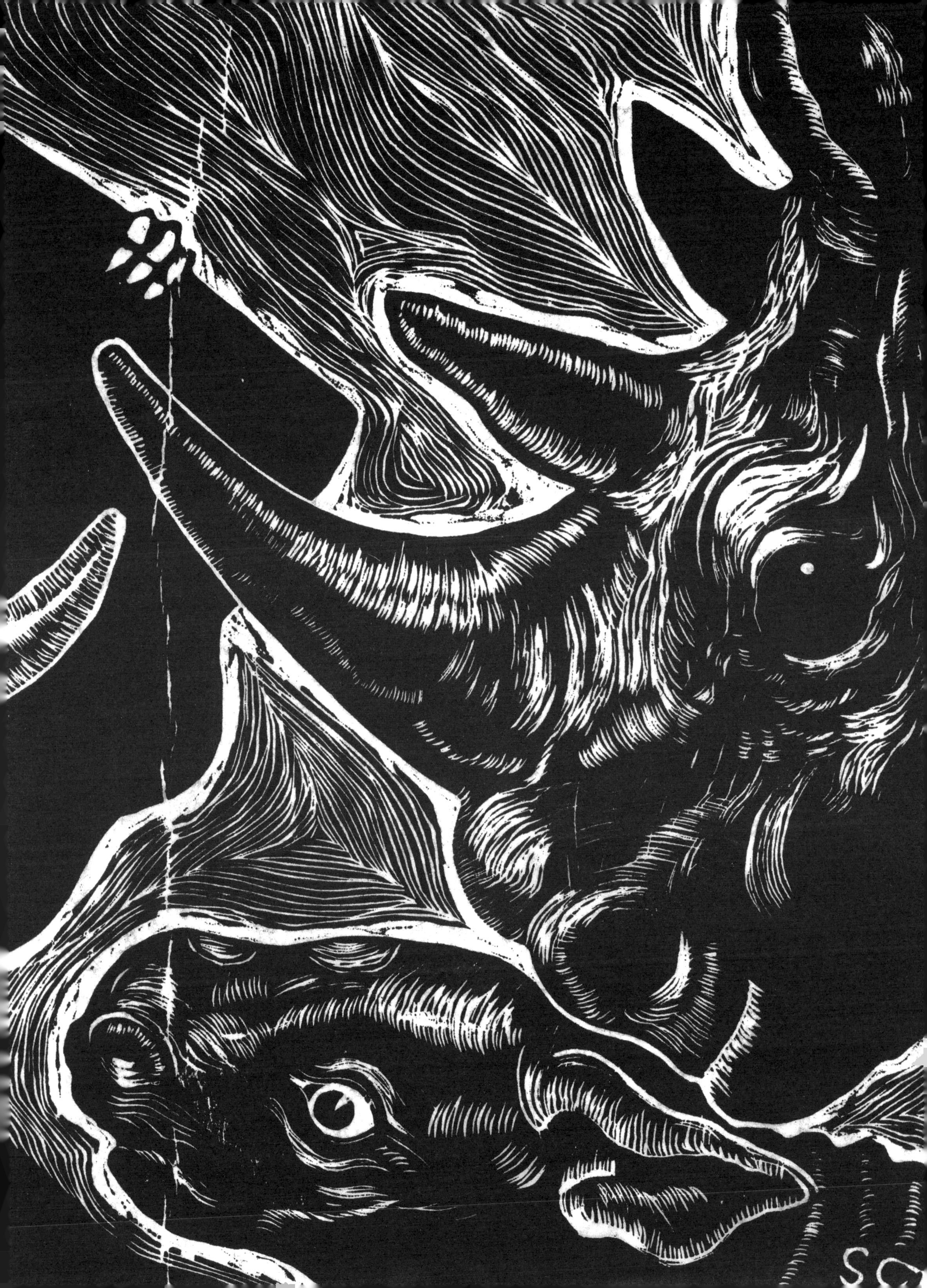

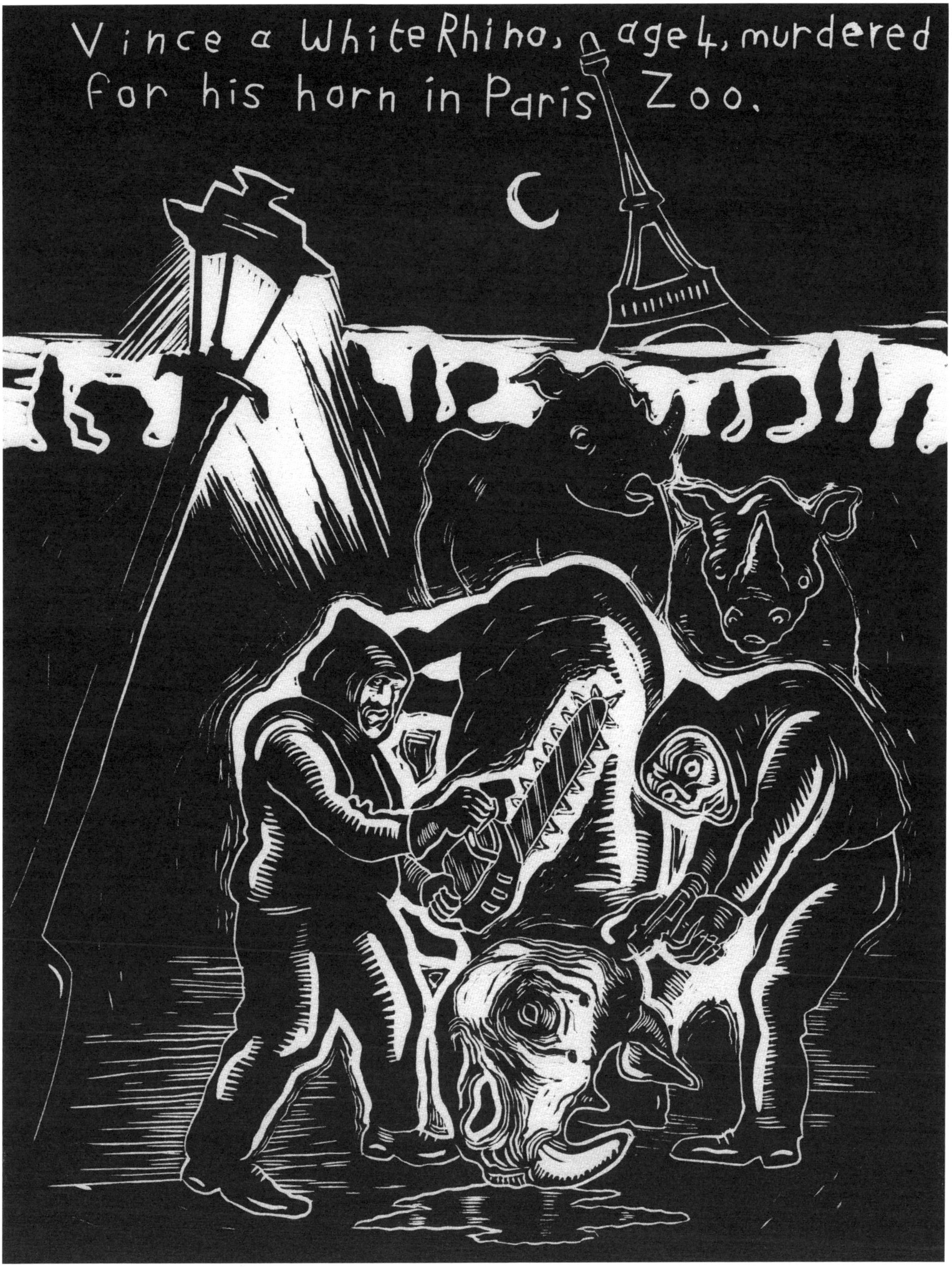
Vince a White Rhino, age 4, murdered
for his horn in Paris Zoo.

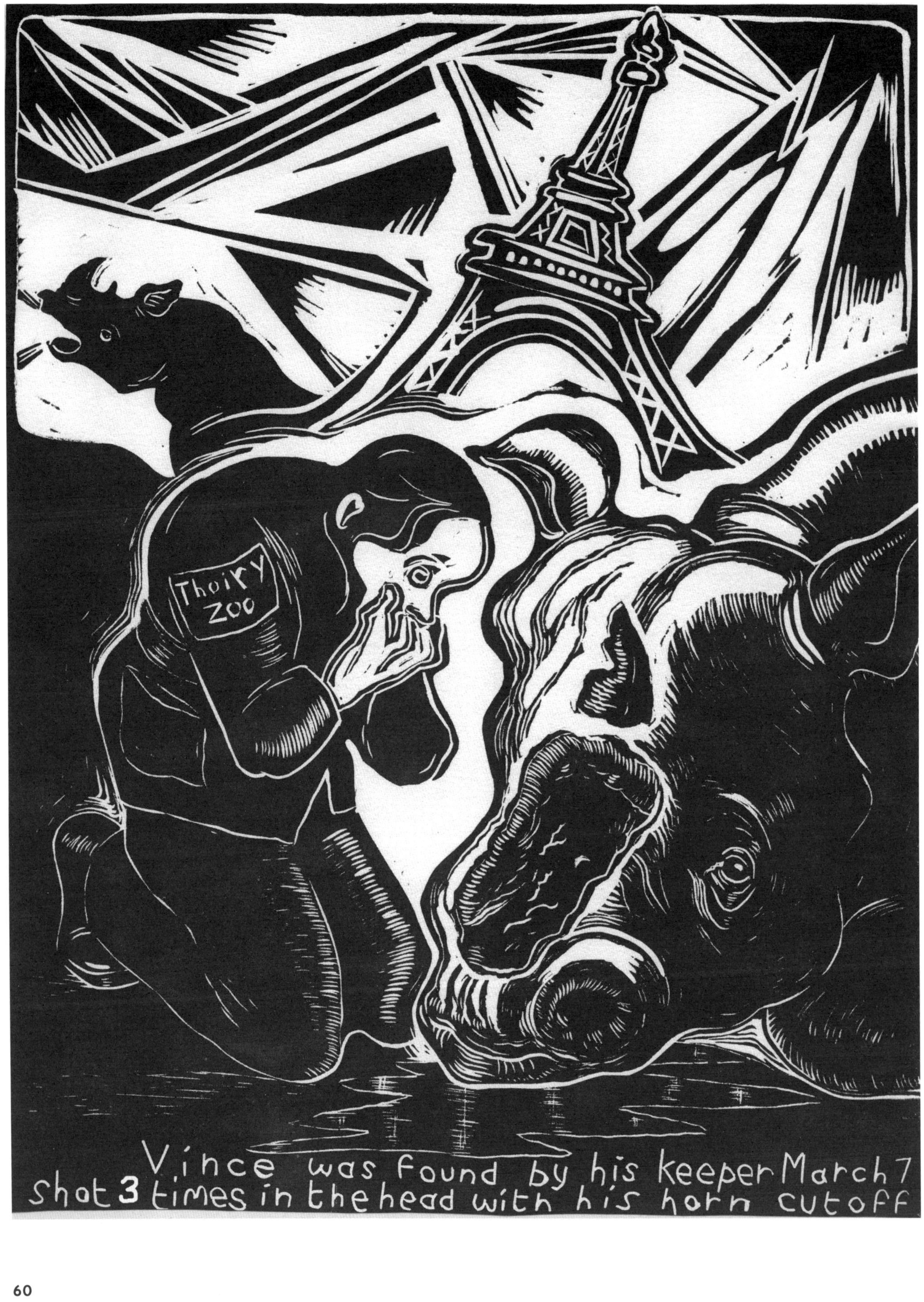
Thoiry Zoo
Vince was found by his keeper March 7
Shot 3 times in the head with his horn cutoff

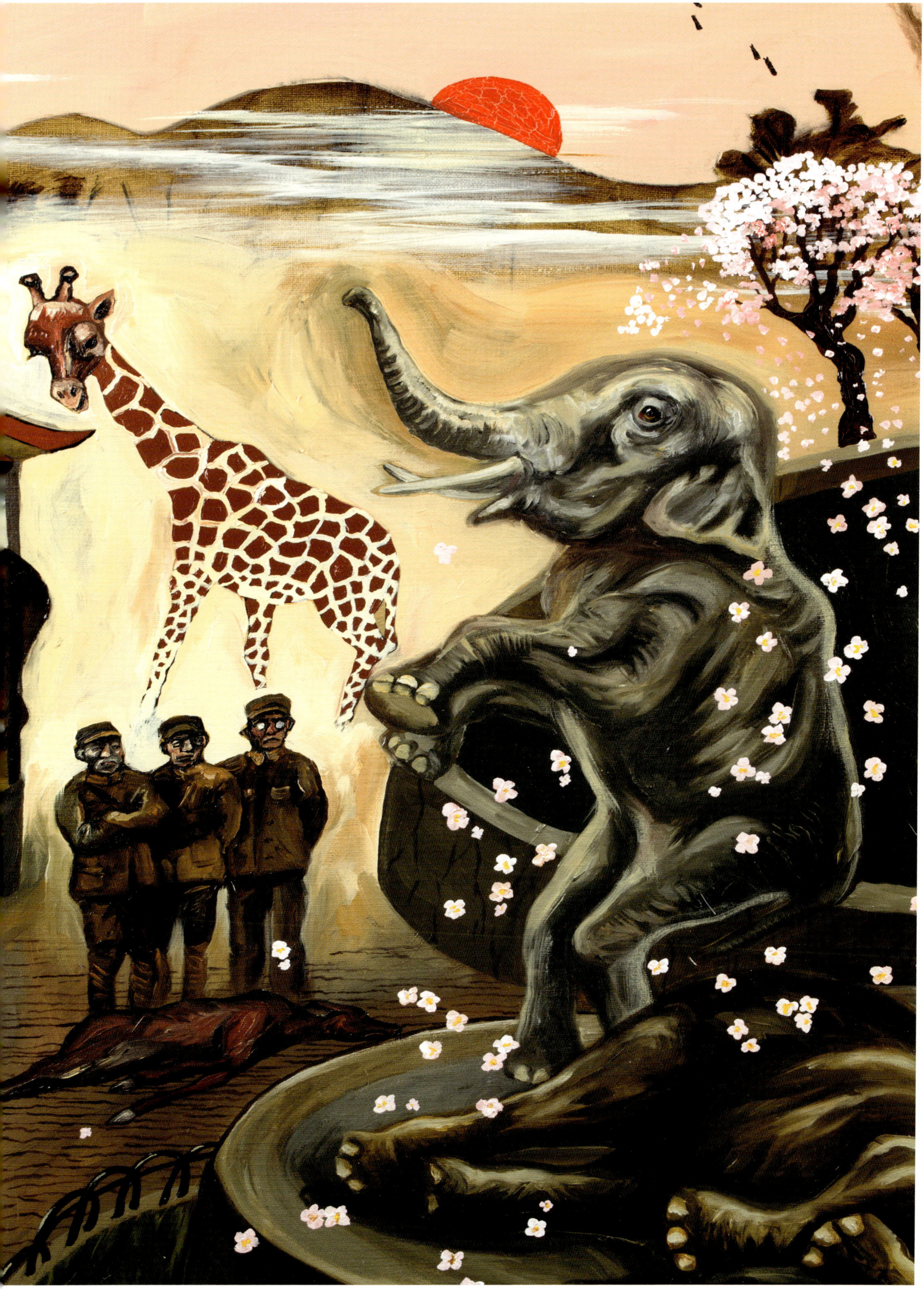

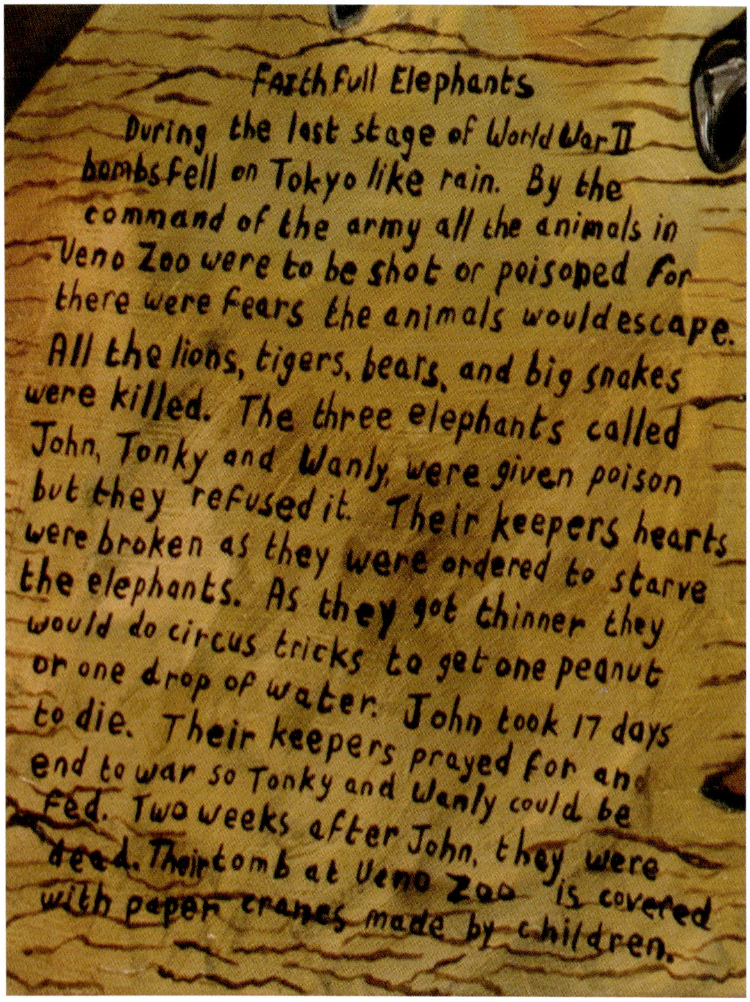
Faithfull Elephants
During the last stage of World War II
bombs fell on Tokyo like rain. By the
command of the army all the animals in
Ueno Zoo were to be shot or poisoned for
there were fears the animals would escape.
All the lions, tigers, bears, and big snakes
were killed. The three elephants called
John, Tonky and Wanly, were given poison
but they refused it. Their keepers hearts
were broken as they were ordered to starve
the elephants. As they got thinner they
would do circus tricks to get one peanut
or one drop of water. John took 17 days
to die. Their keepers prayed for an
end to war so Tonky and Wanly could be
fed. Two weeks after John, they were
dead. Their tomb at Ueno Zoo is covered
with paper cranes made by children.

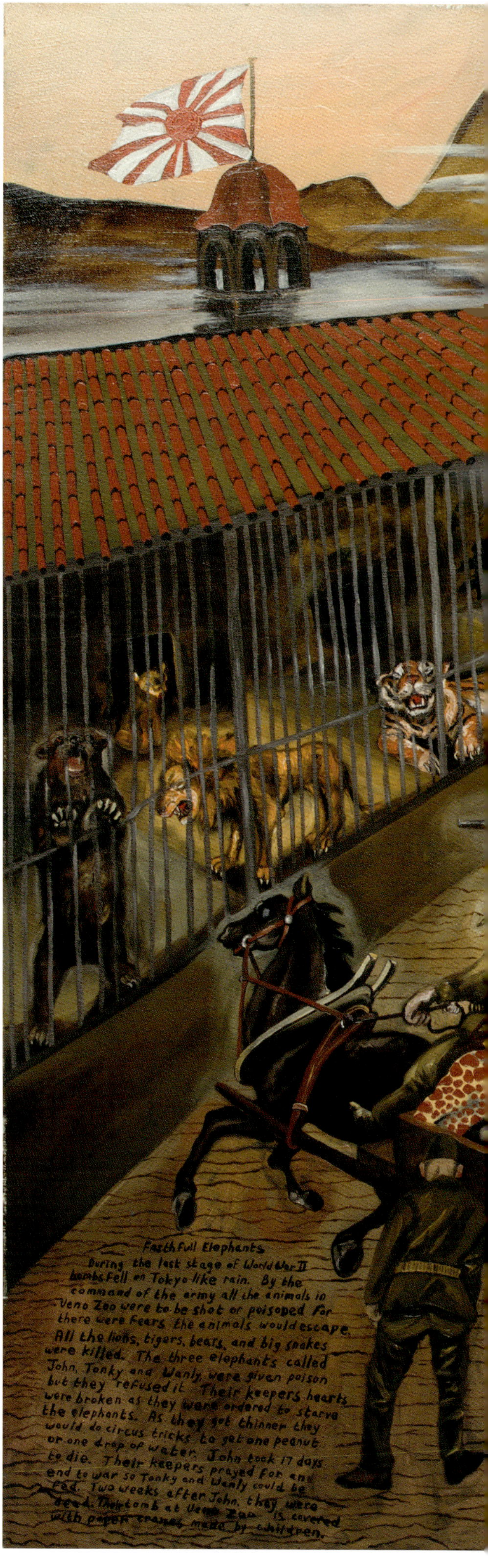

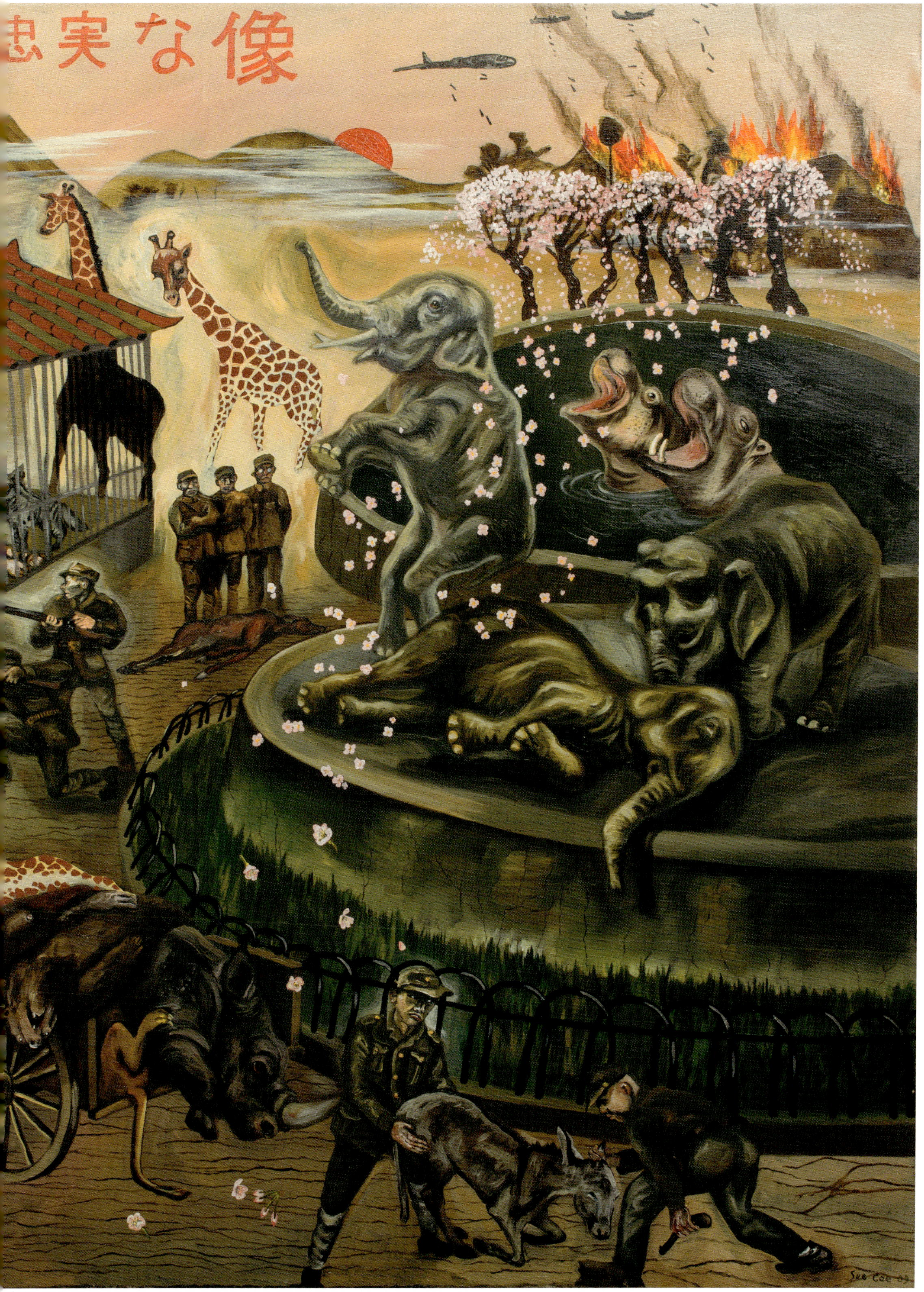

実な像
Sue Coe 03

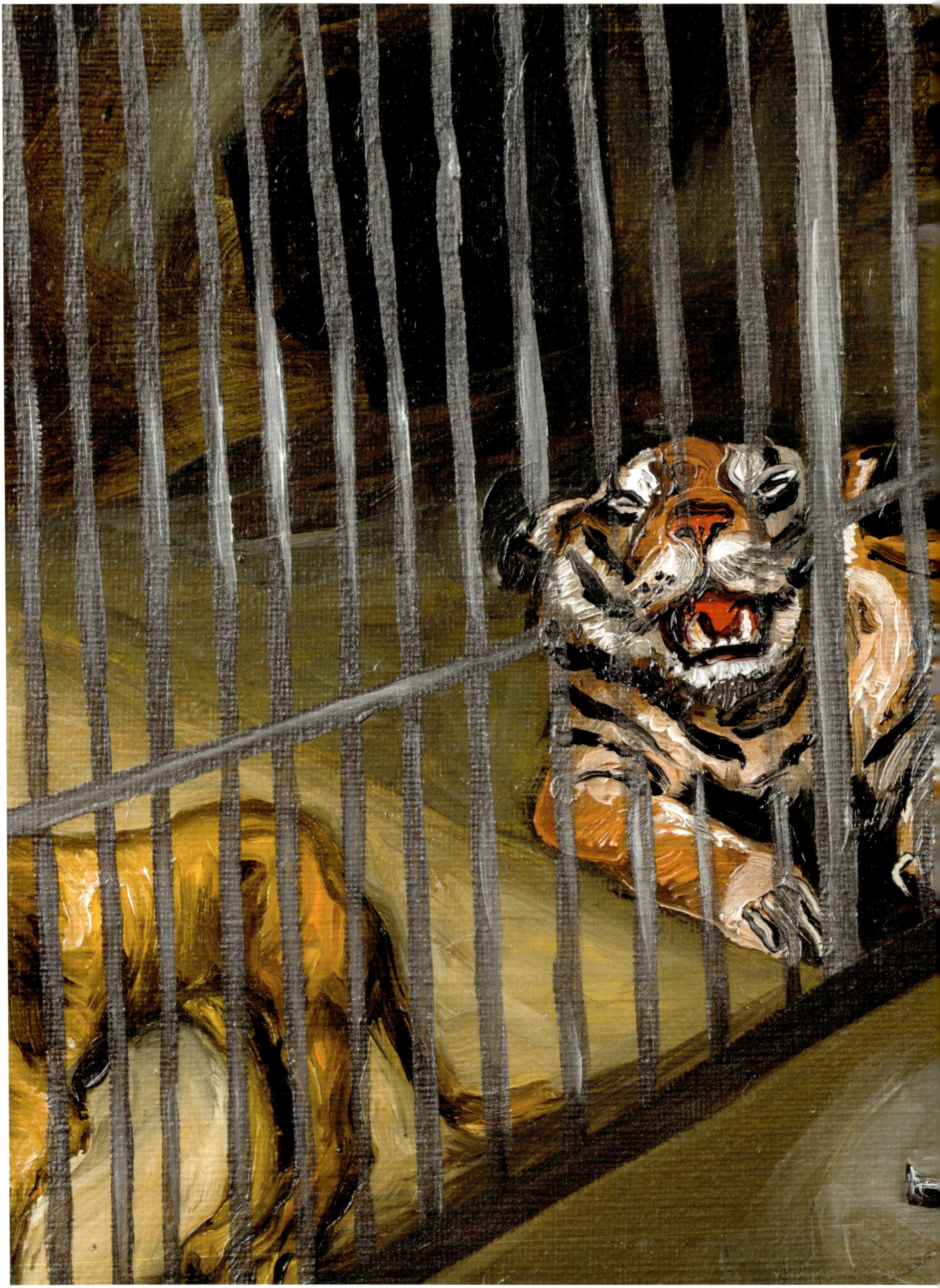

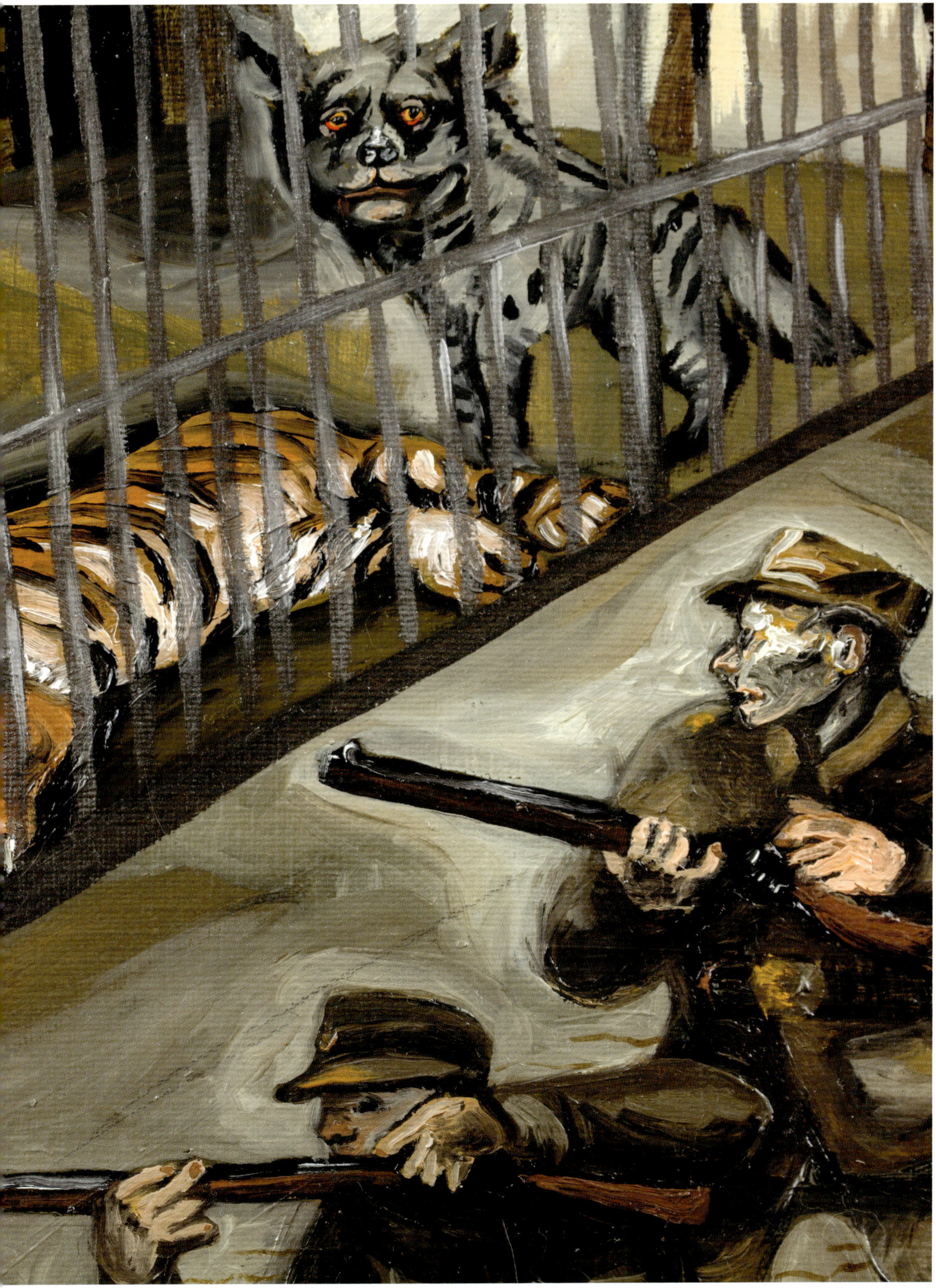

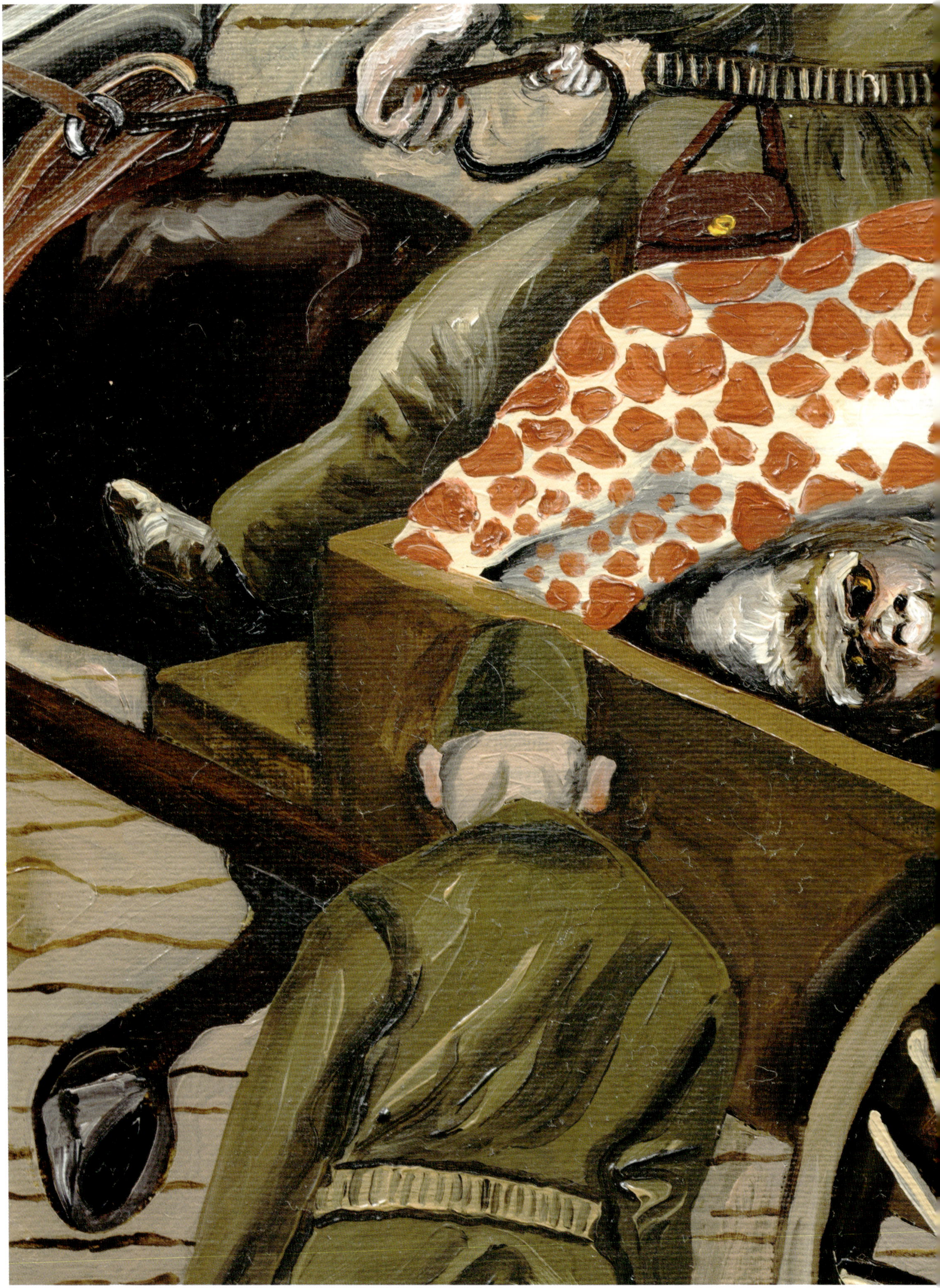

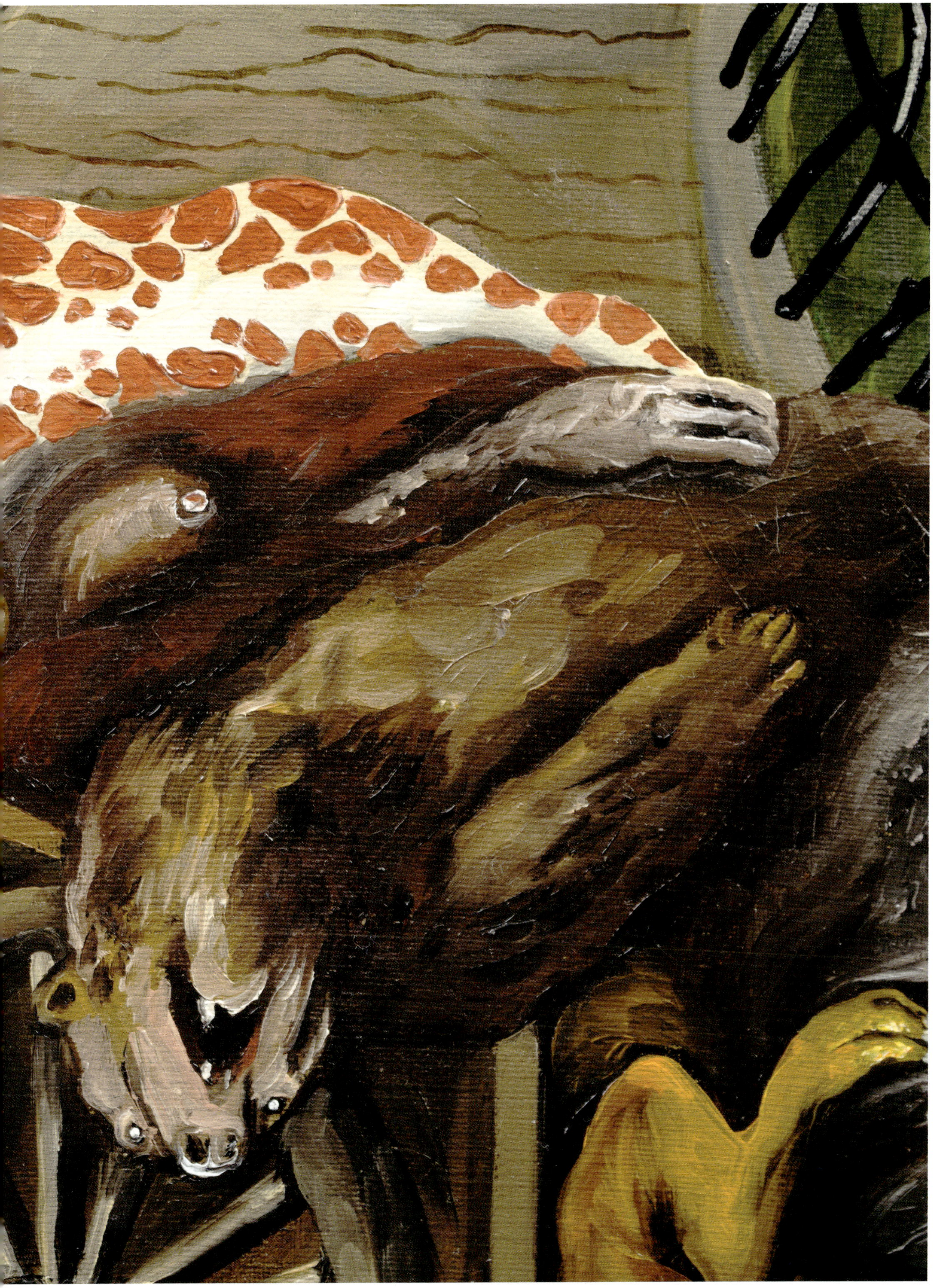

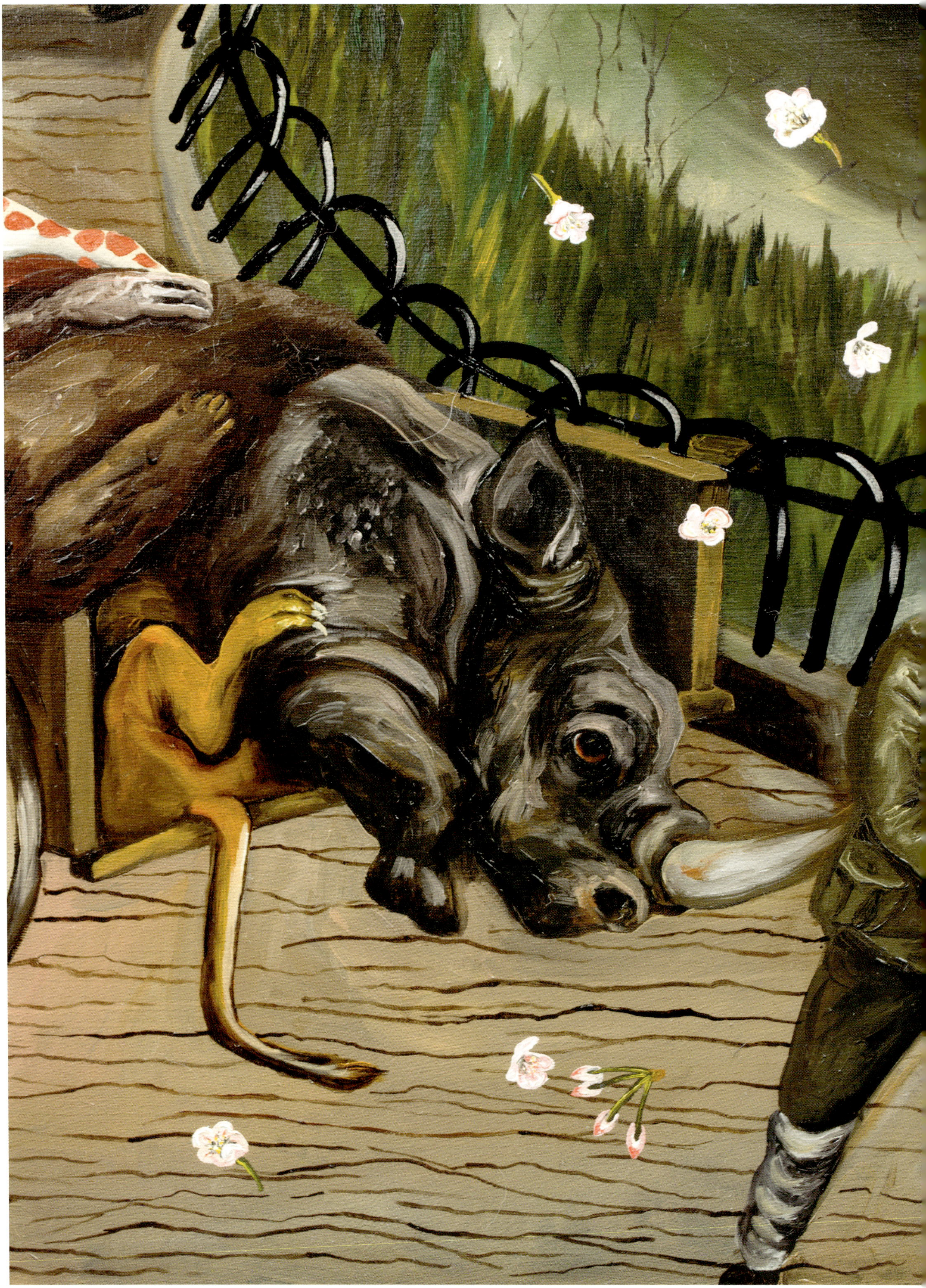

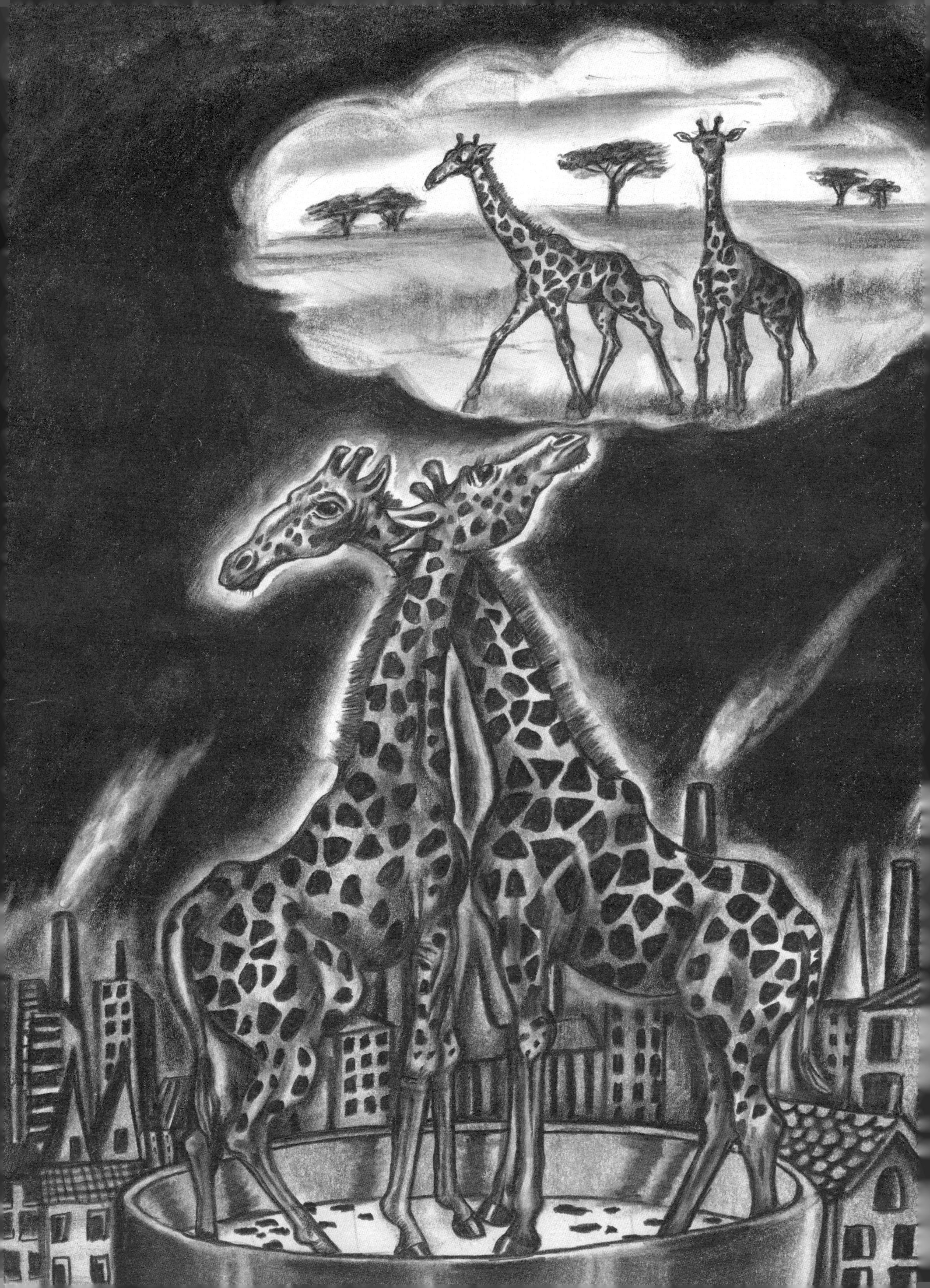

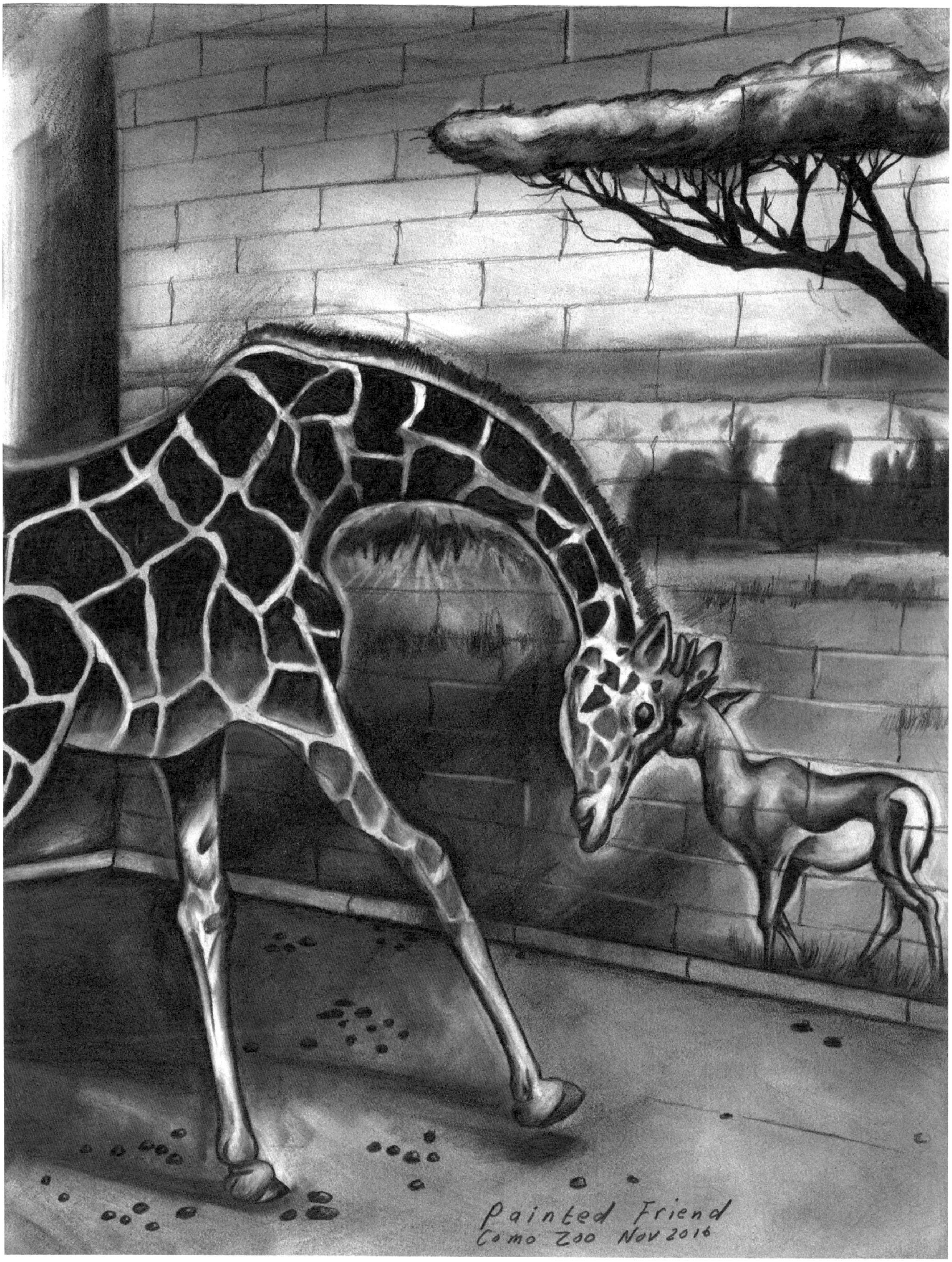
Painted Friend
Como Zoo Nov 2016

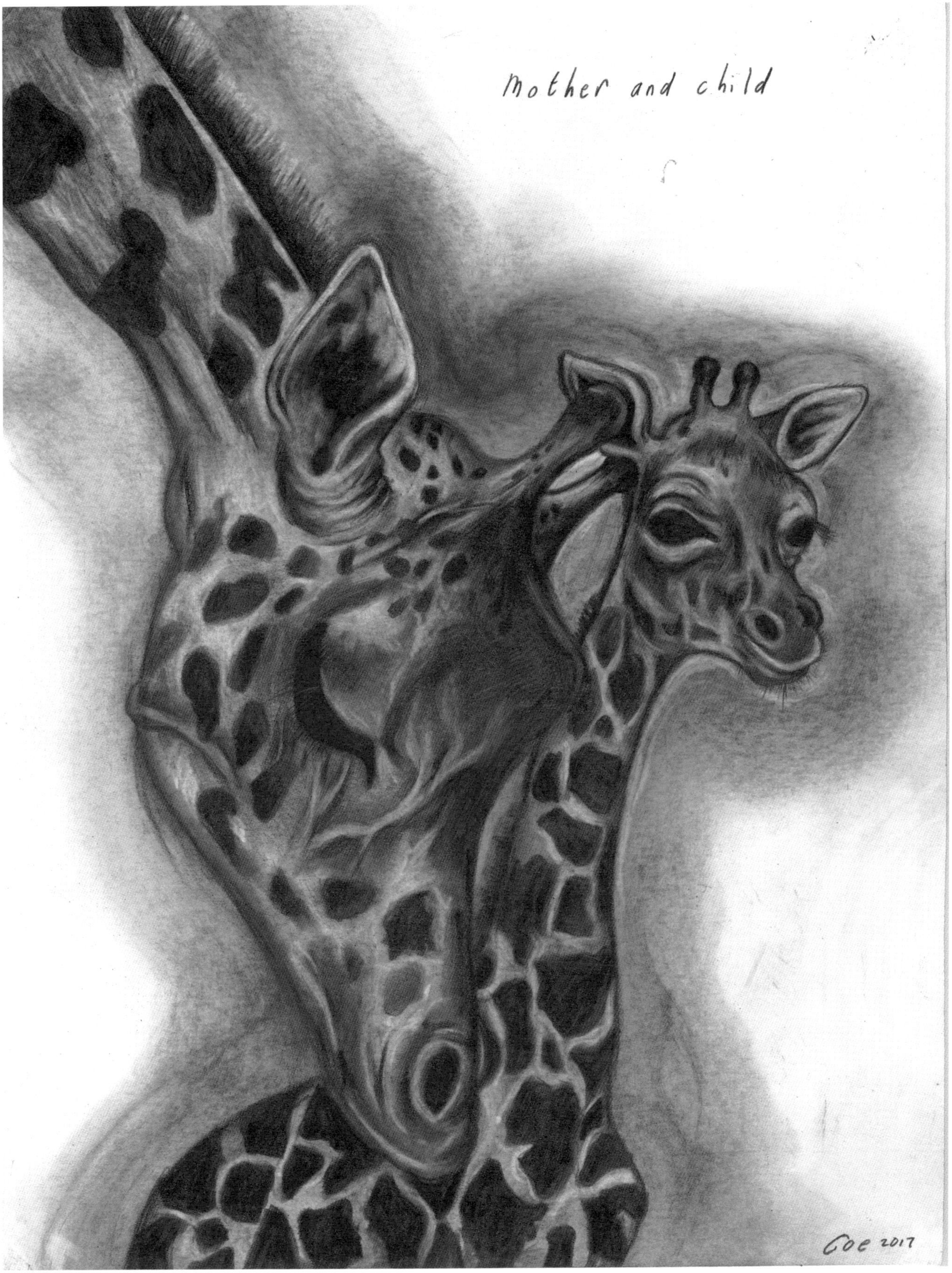
Mother and child
Coe 2017

Copenhagen Zoo Feb 2014
Marius, a healthy 18 month old giraffe, was deemed 'genetic surplus' and shot in the head.

Then Marius was butchered in front of school children. His body fed to the lions. A month later, the lions and their two 10 month old cubs were killed to make room for a younger lion, whom the zoo wanted to breed.

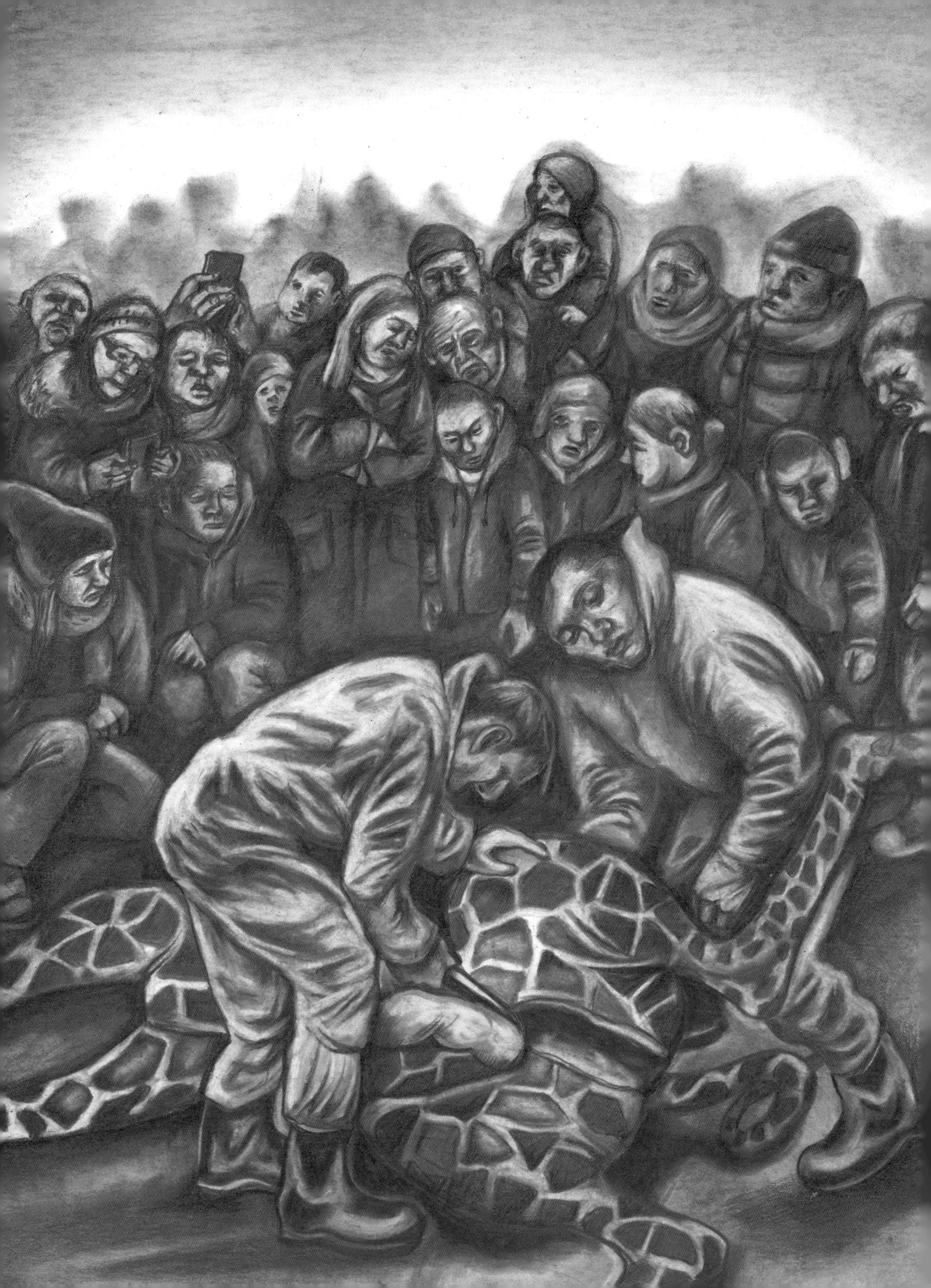

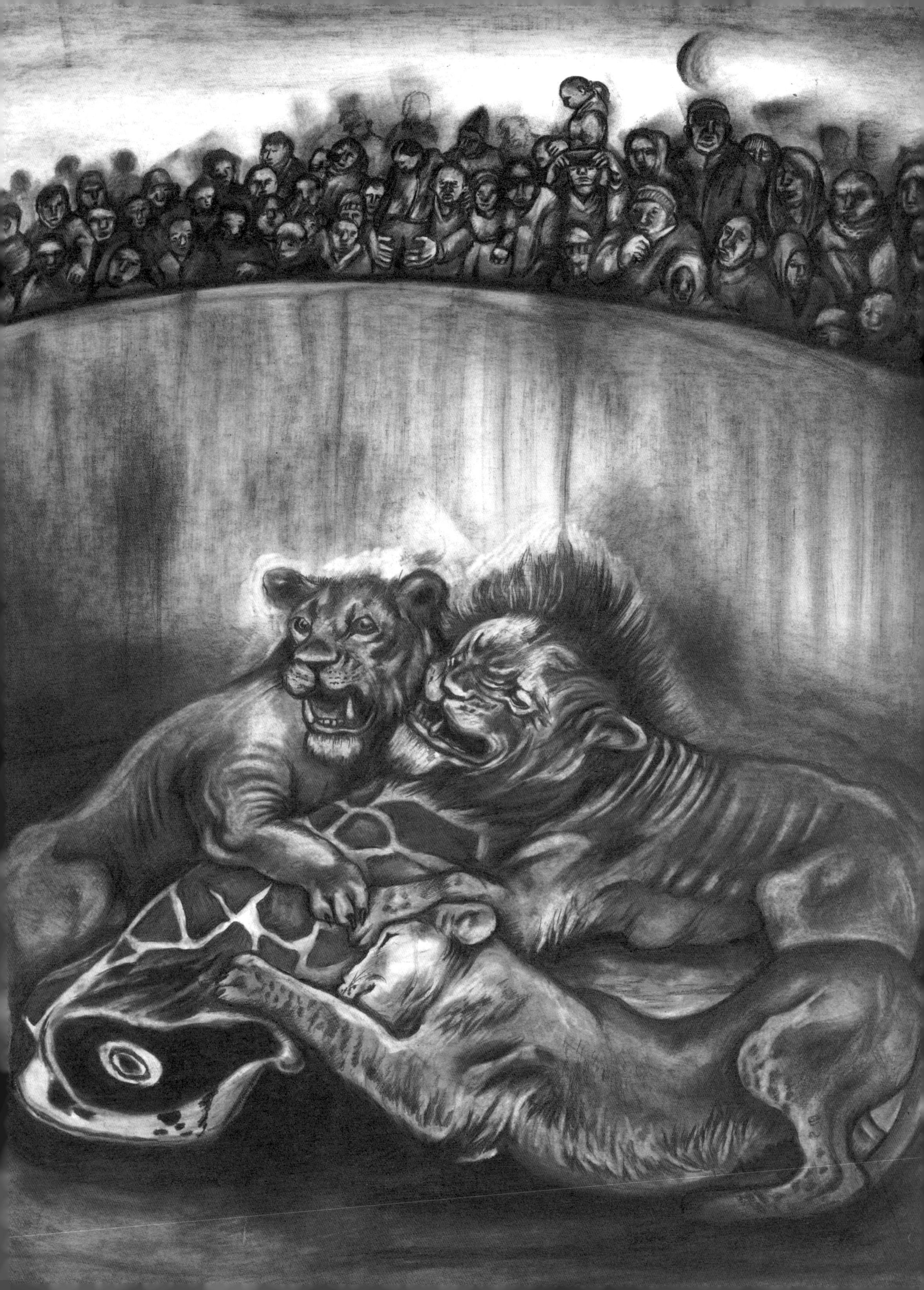

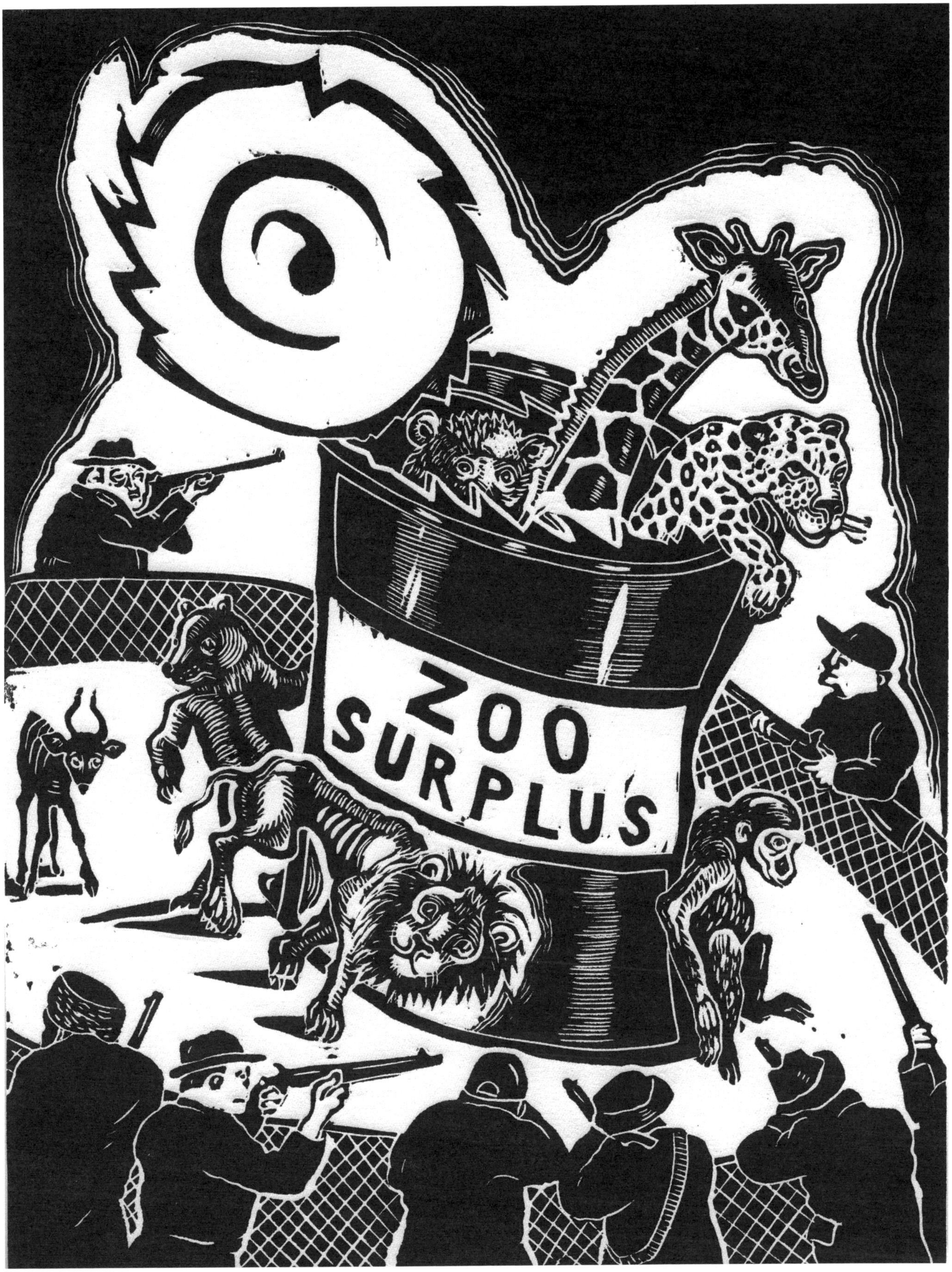
ZOO
SURPLUS

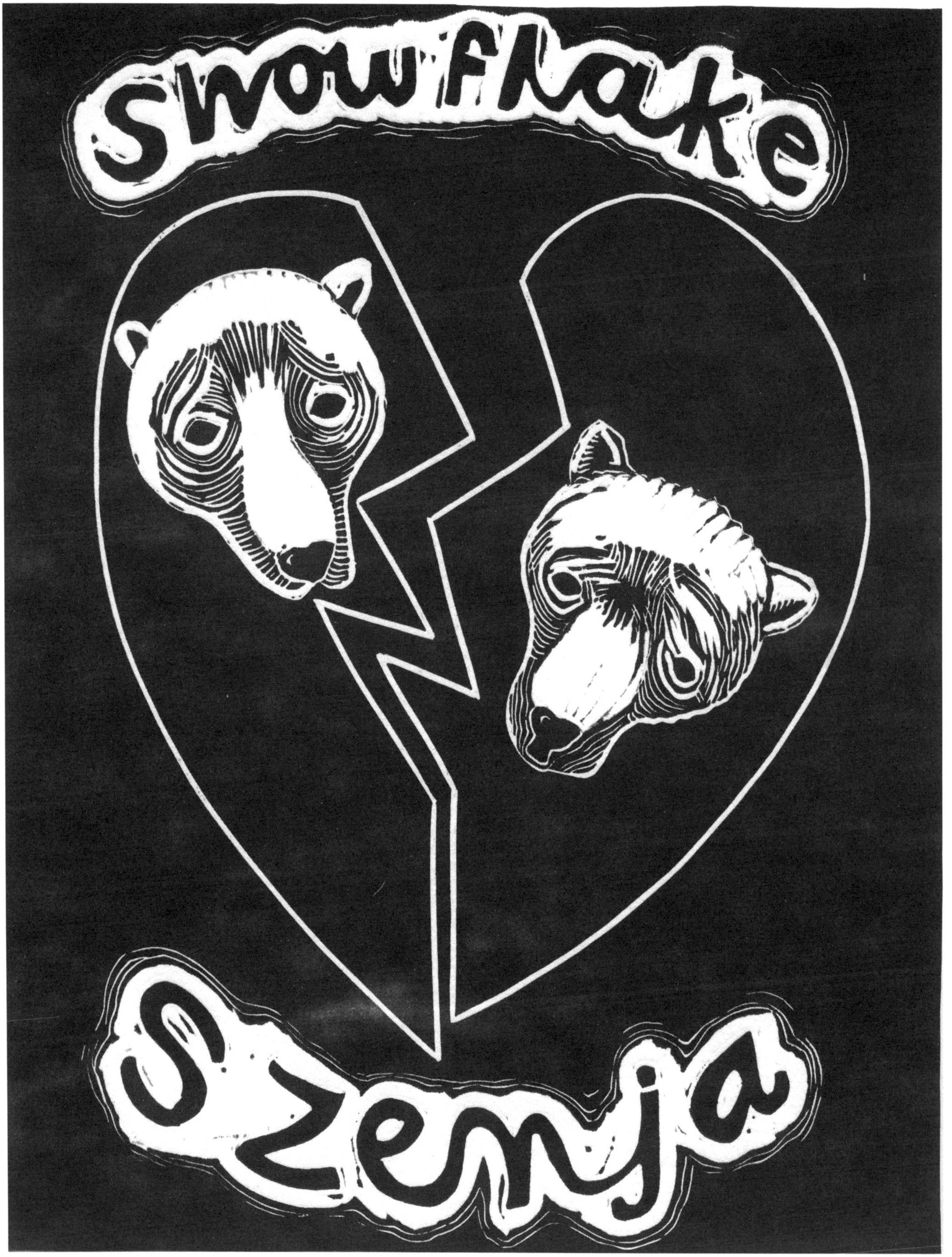
Snowflake
Szenja

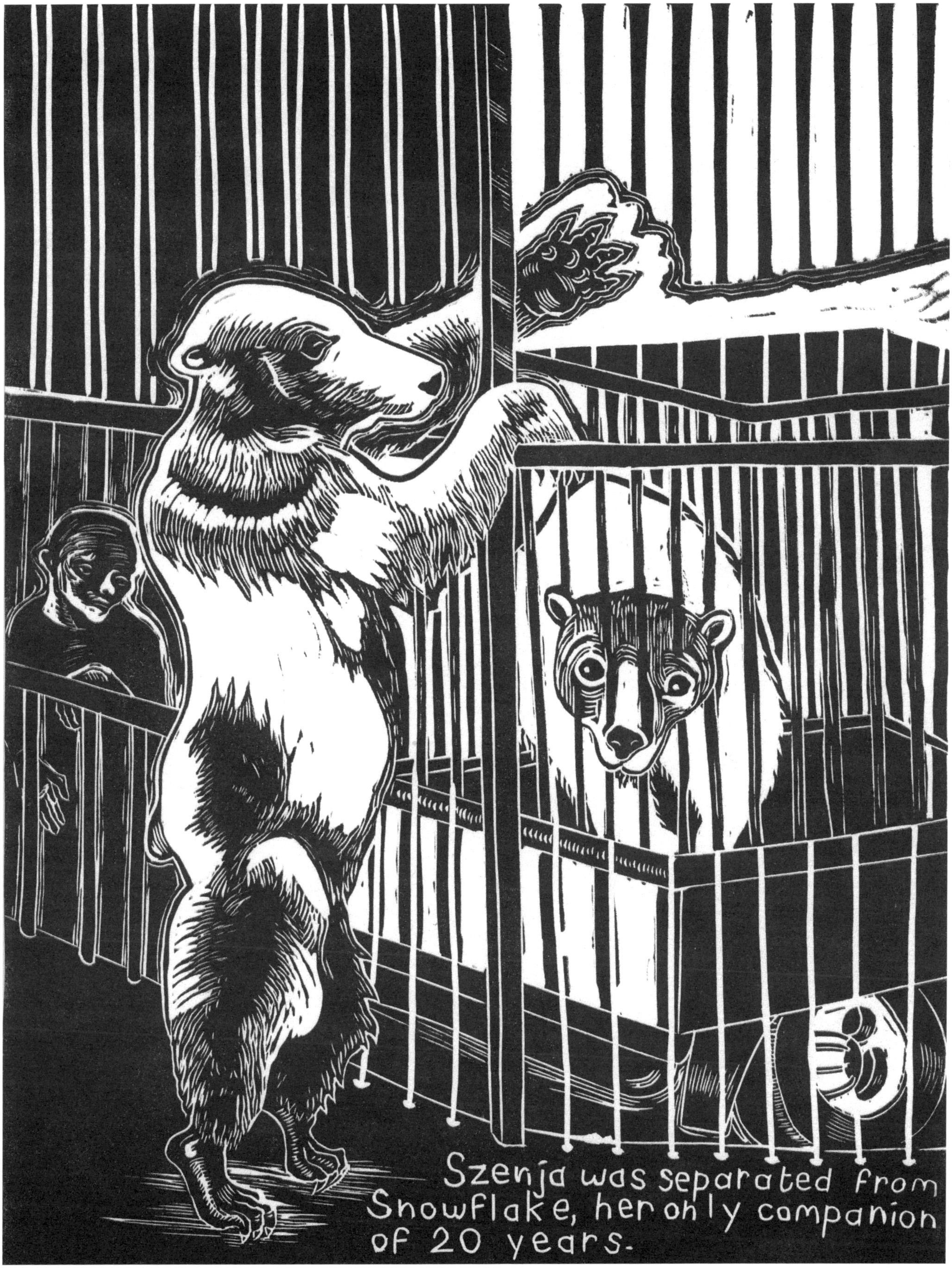
Szenja was separated from
Snowflake, her only companion
of 20 years.

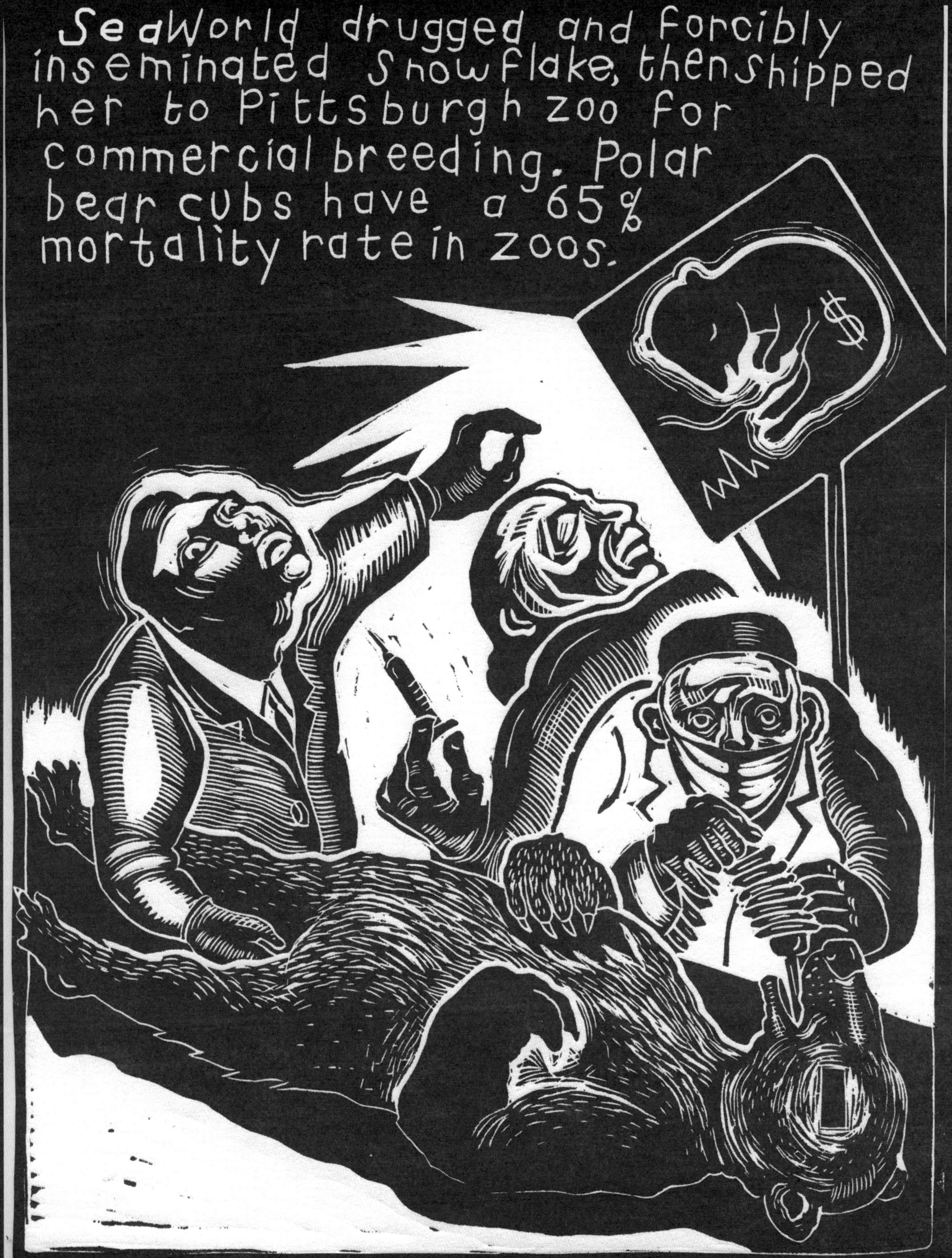
SeaWorld drugged and forcibly inseminated Snowflake, then shipped her to Pittsburgh zoo for commercial breeding. Polar bear cubs have a 65% mortality rate in zoos.
$

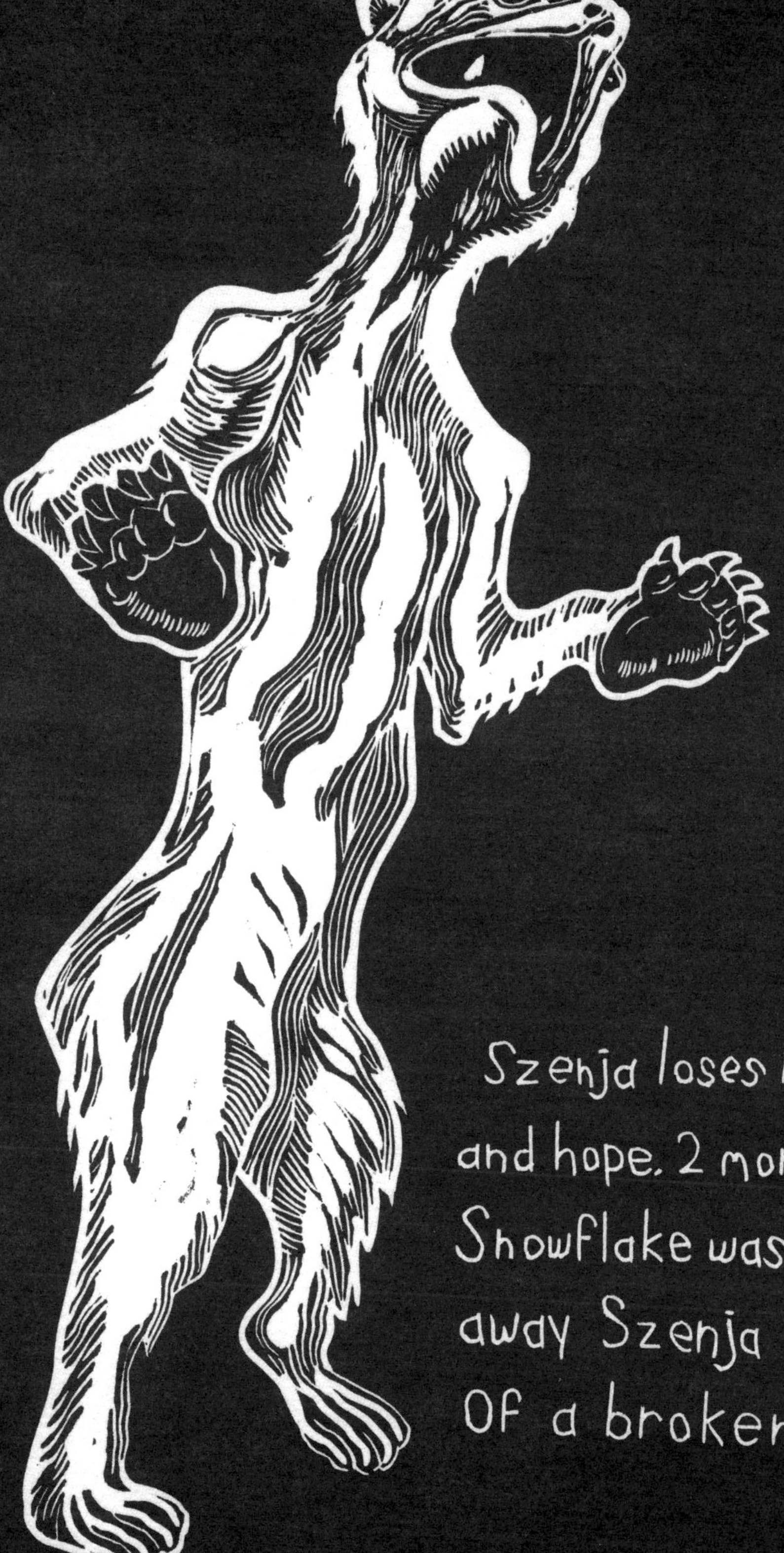

Szenja loses her appetite and hope. 2 months after Snowflake was taken away Szenja died. Of a broken heart?

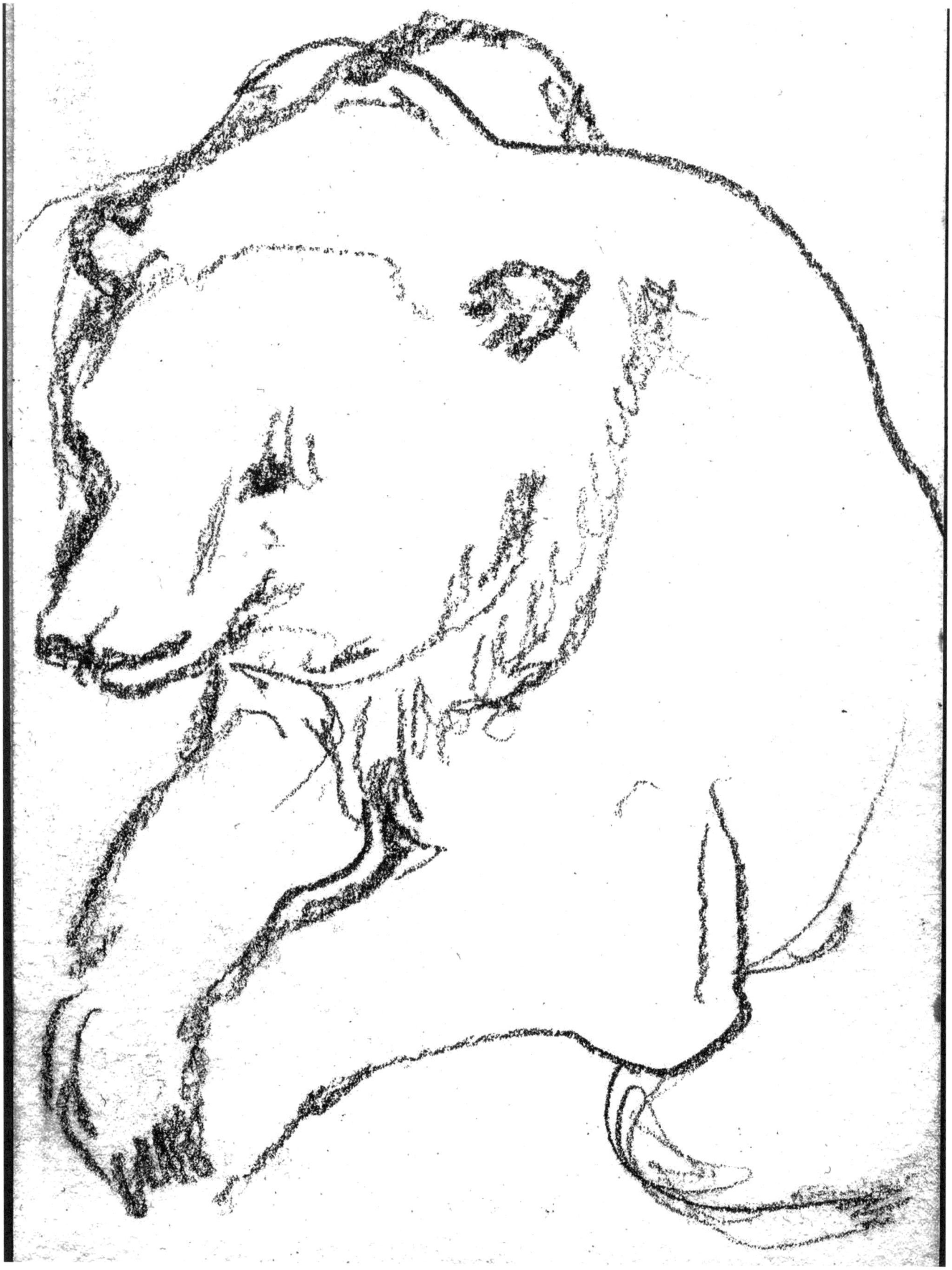

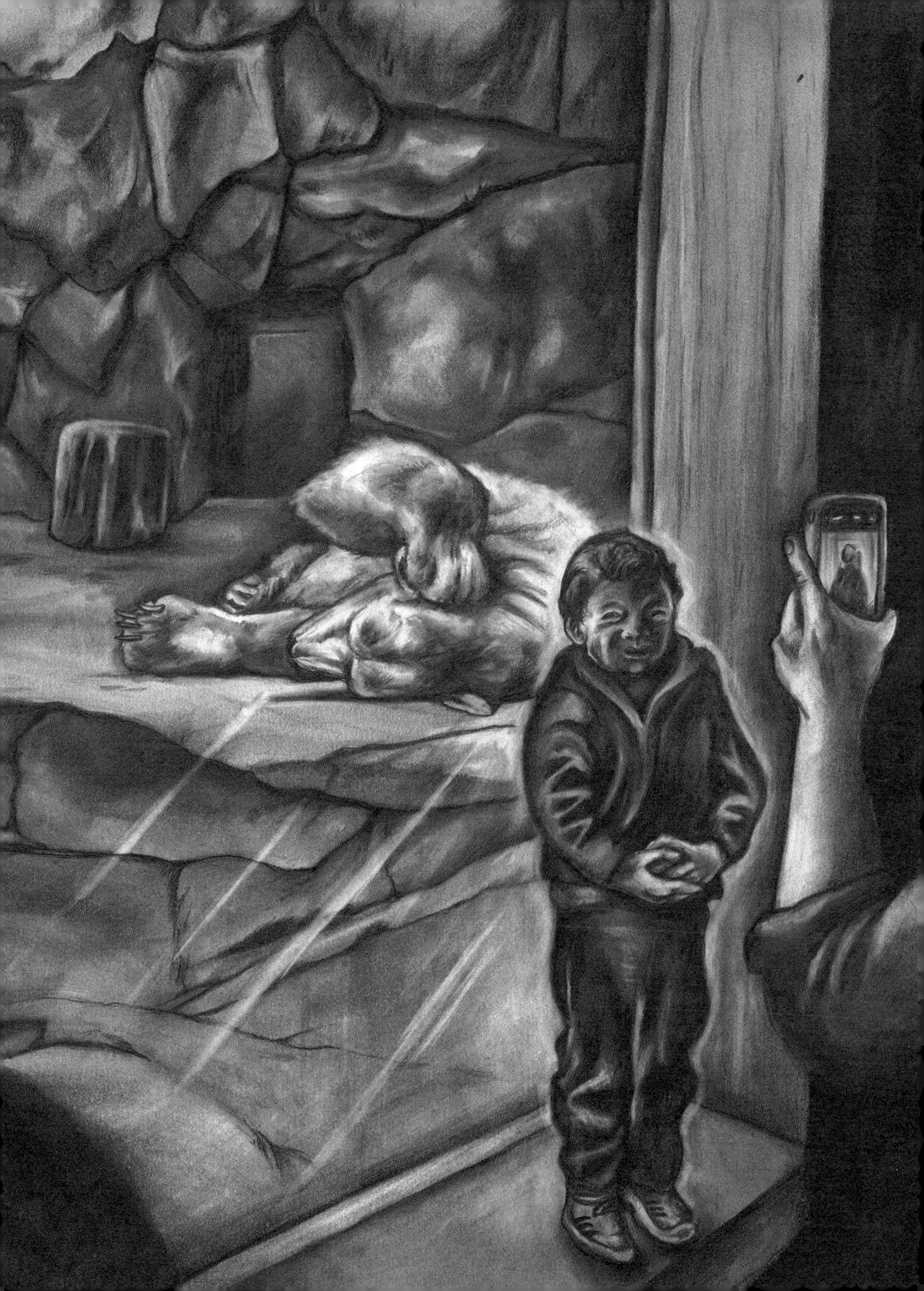

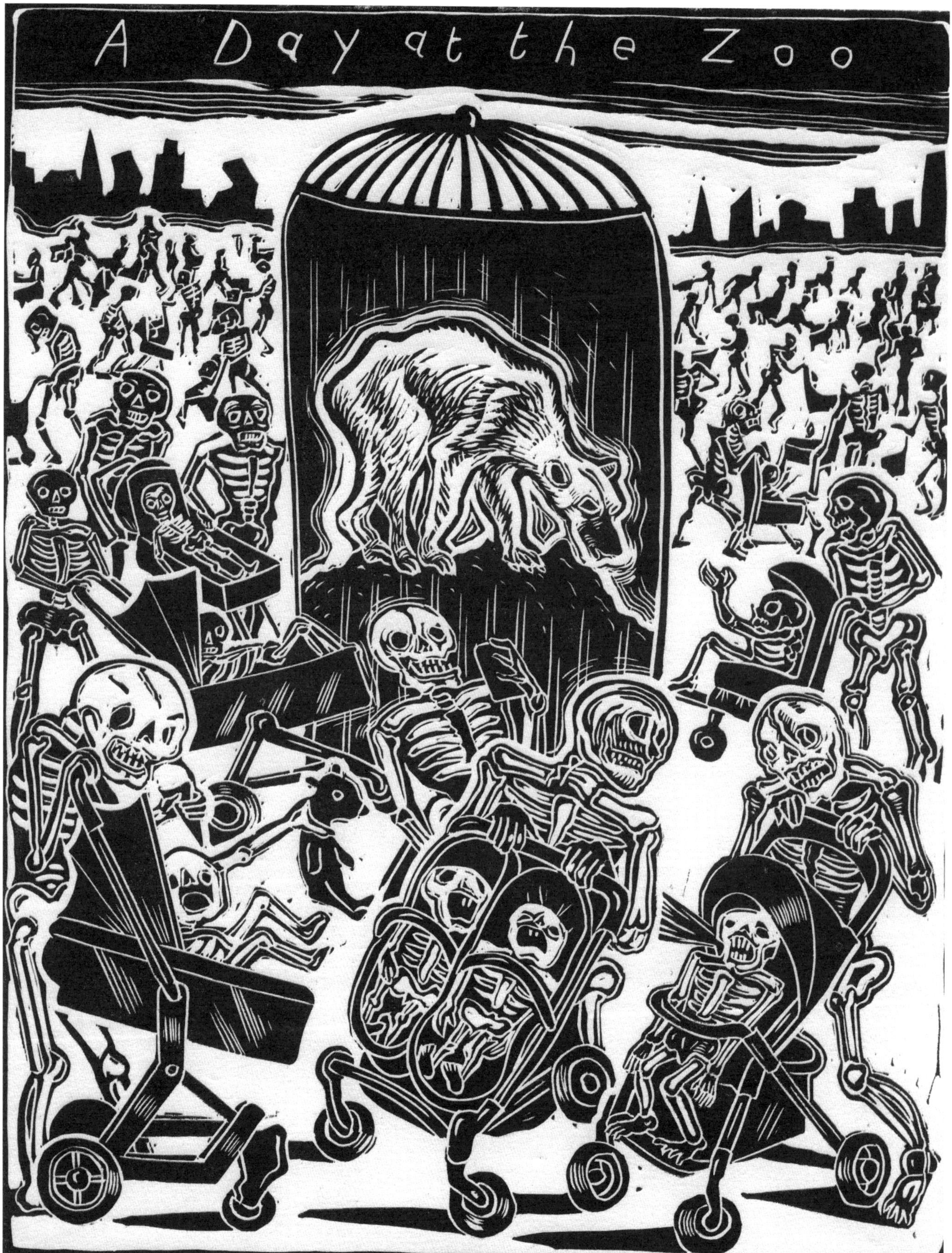
A Day at the Zoo

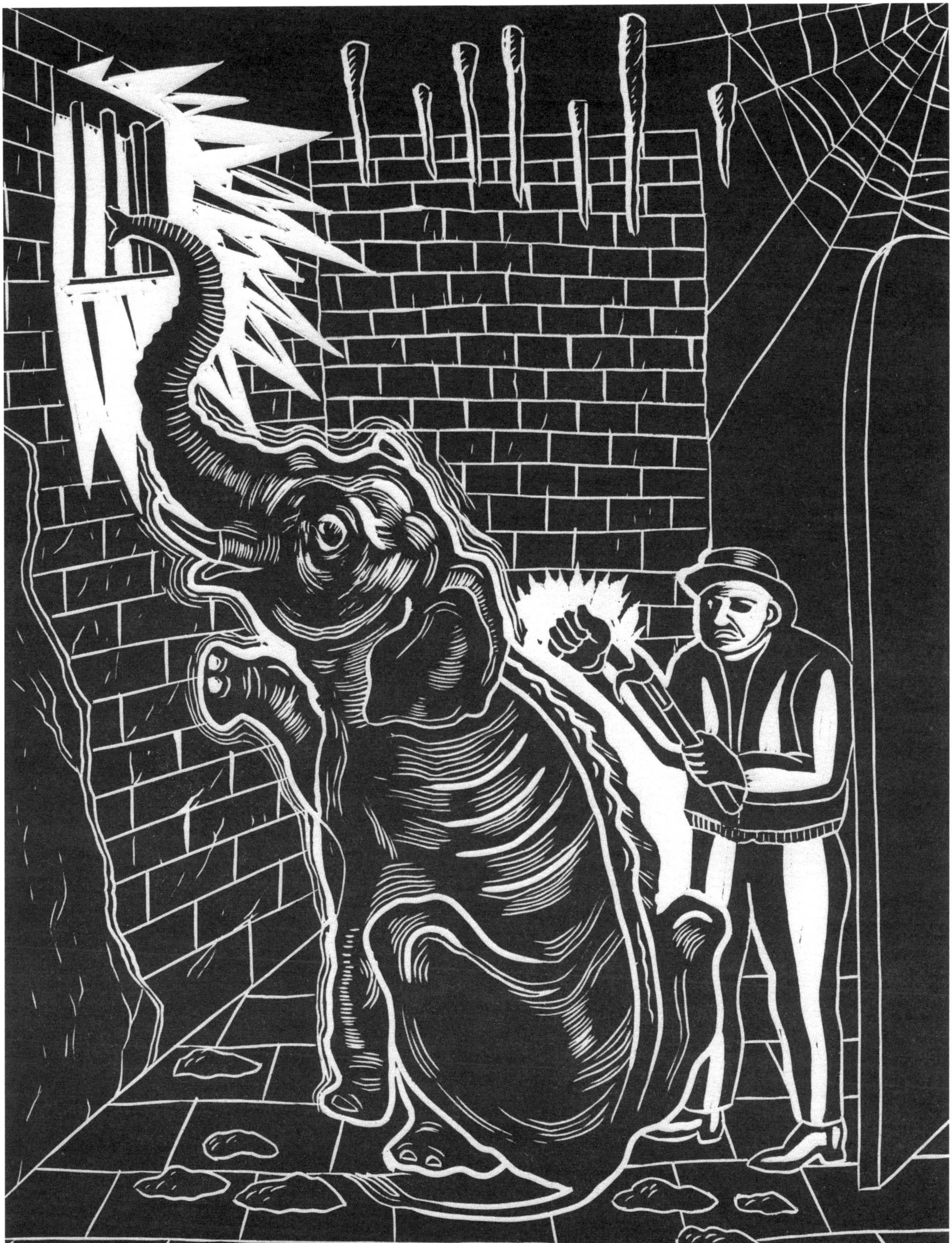

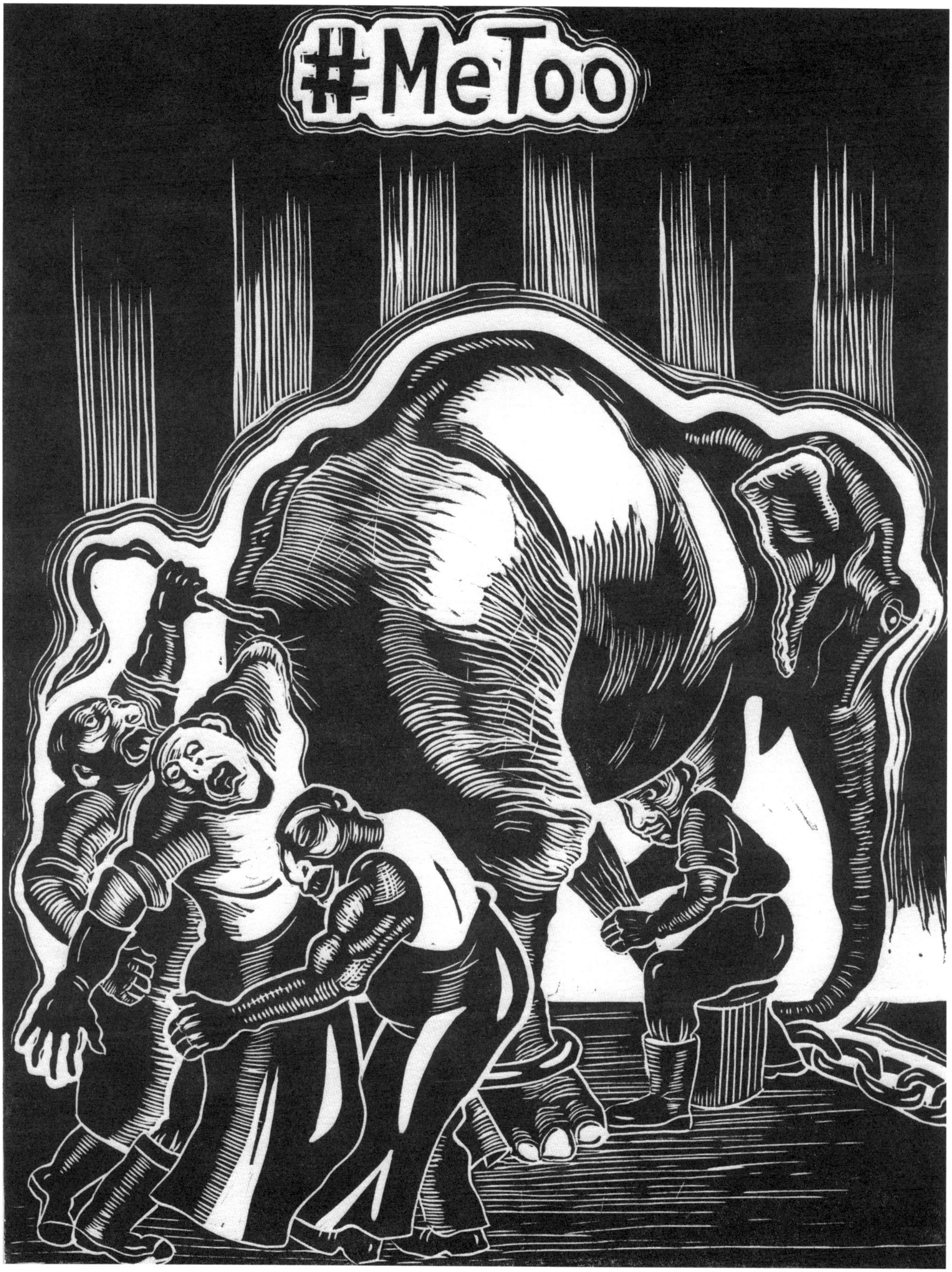
#MeToo

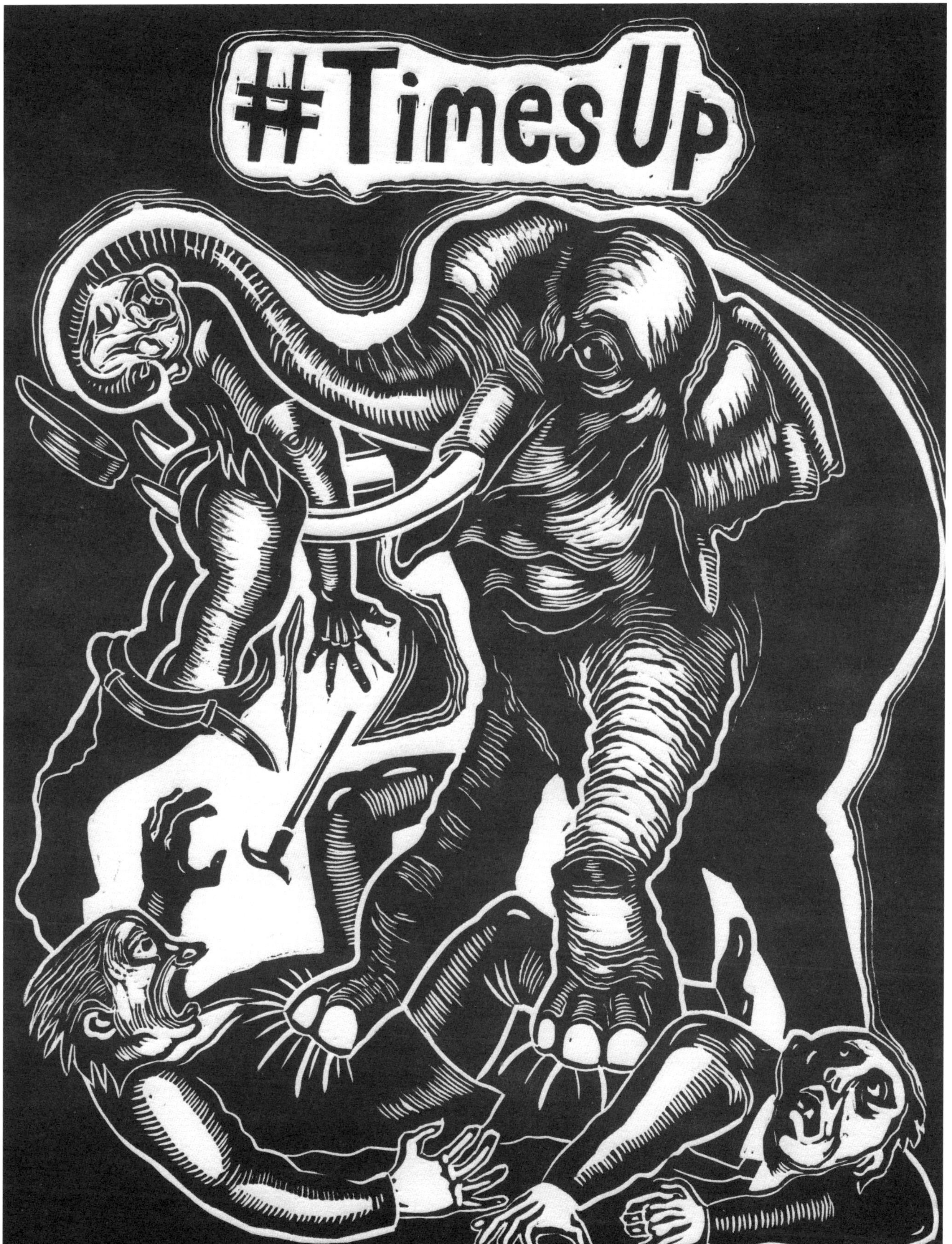
#TimesUp

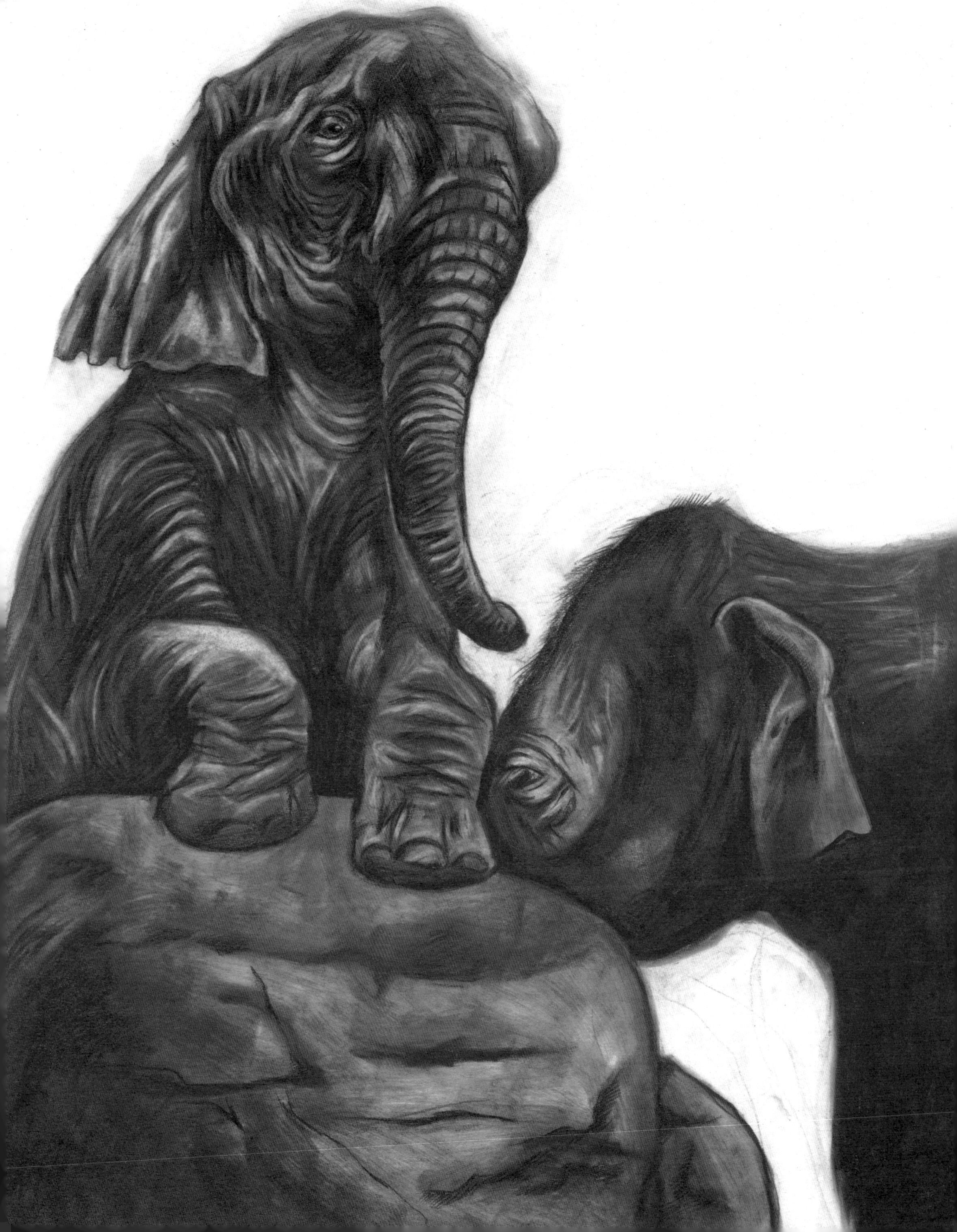

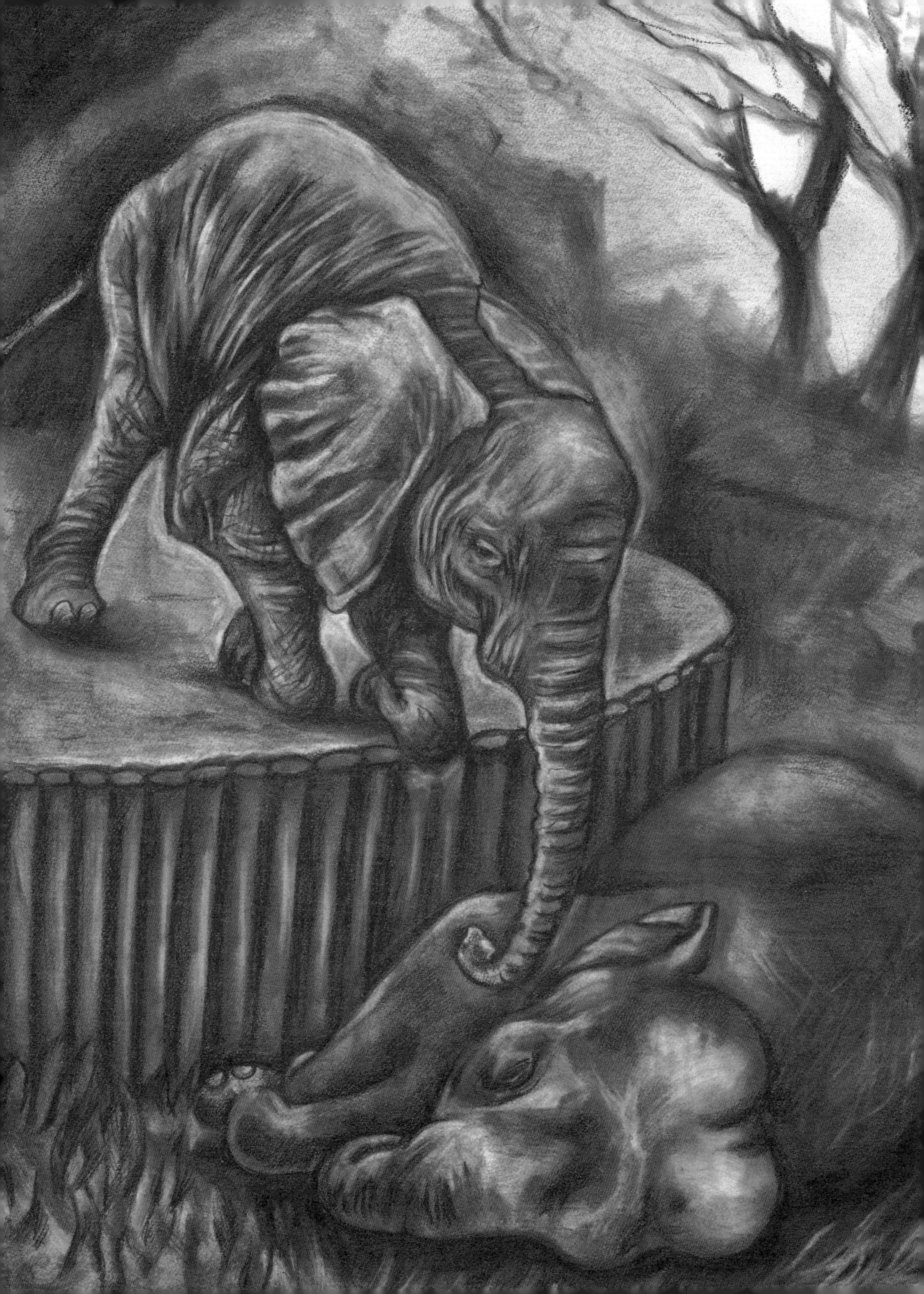

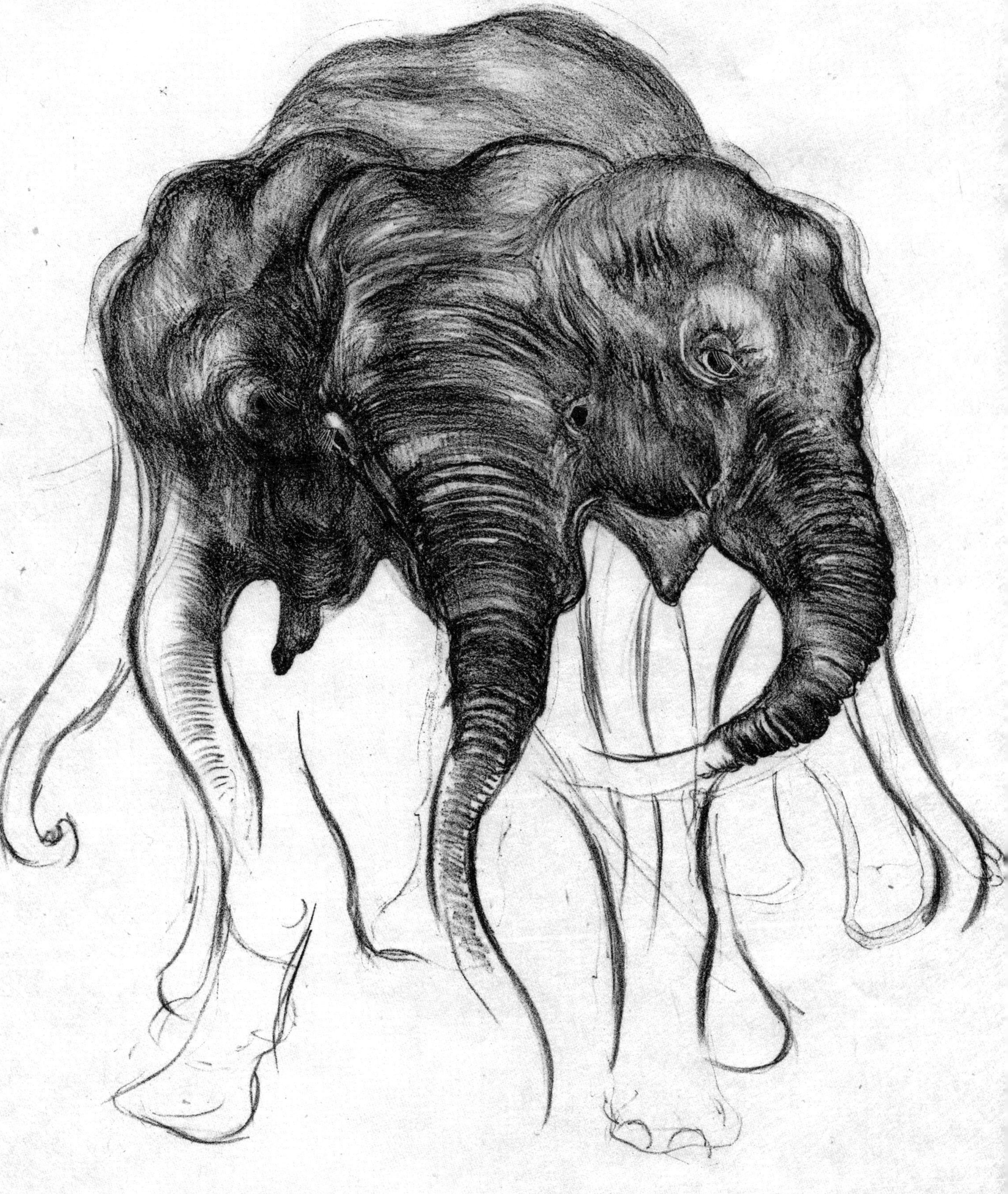

Stress induced stereotypical head swaying

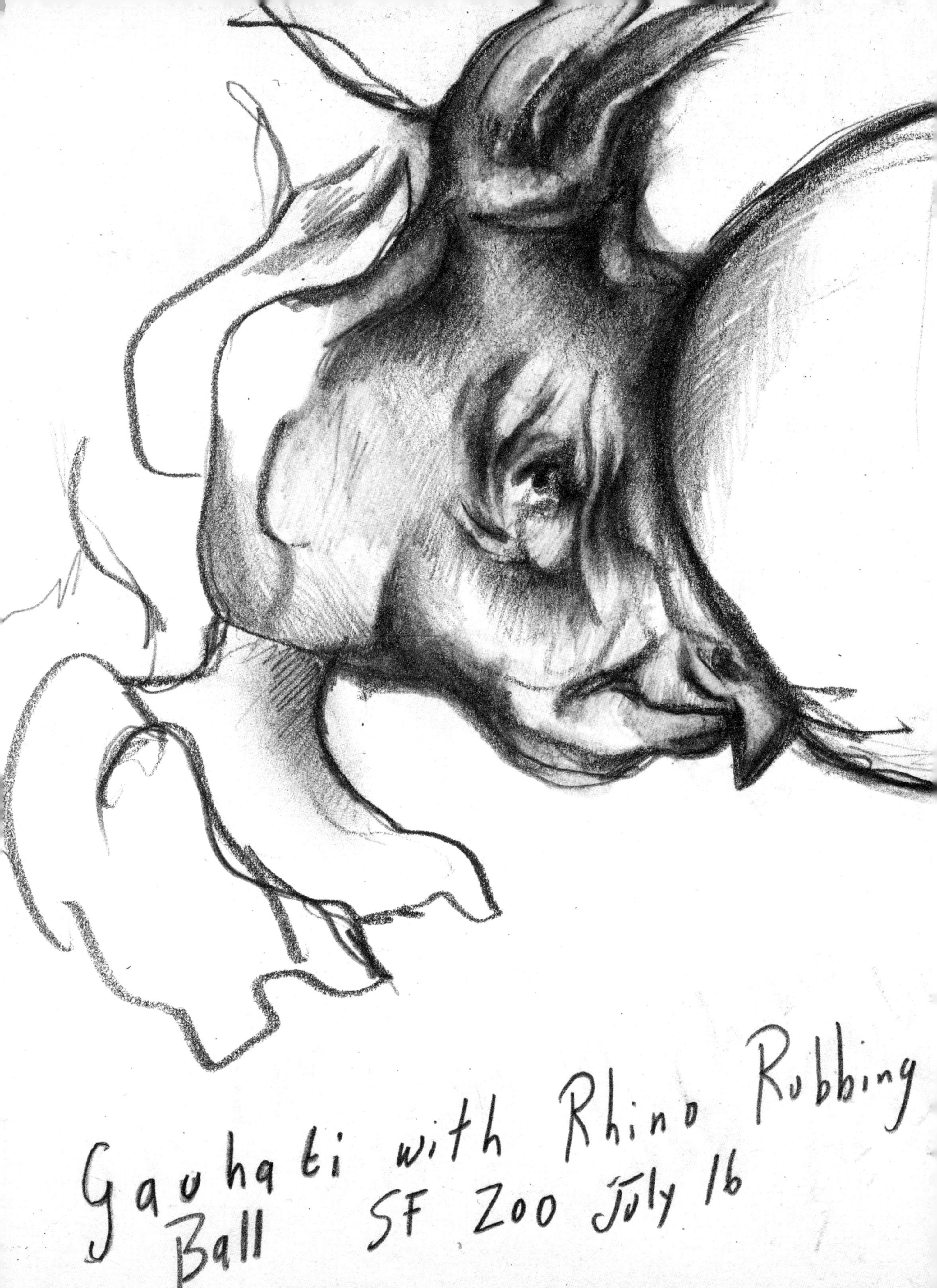

Gauhati with Rhino Rubbing Ball SF Zoo July 16

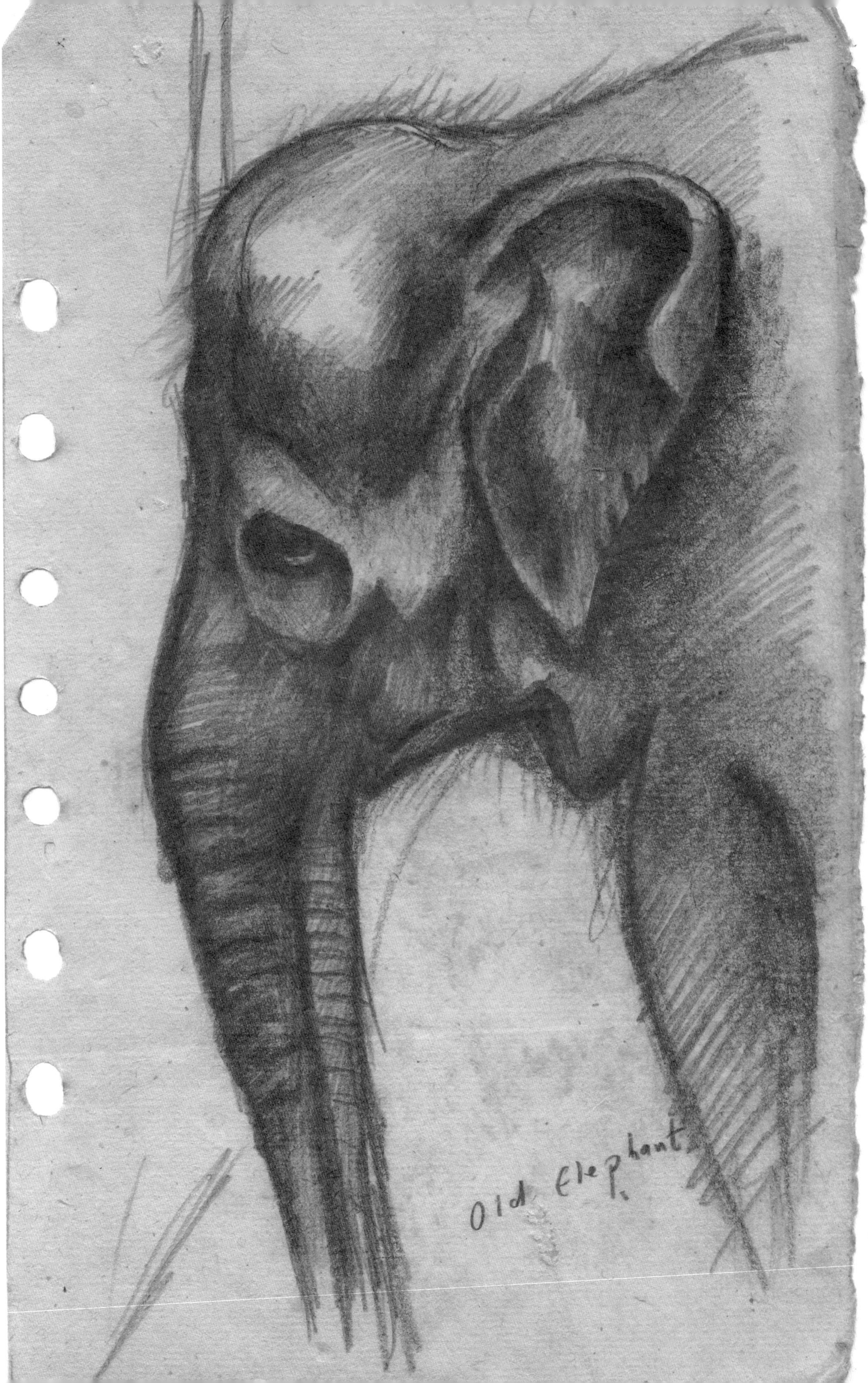
Old Elephant.

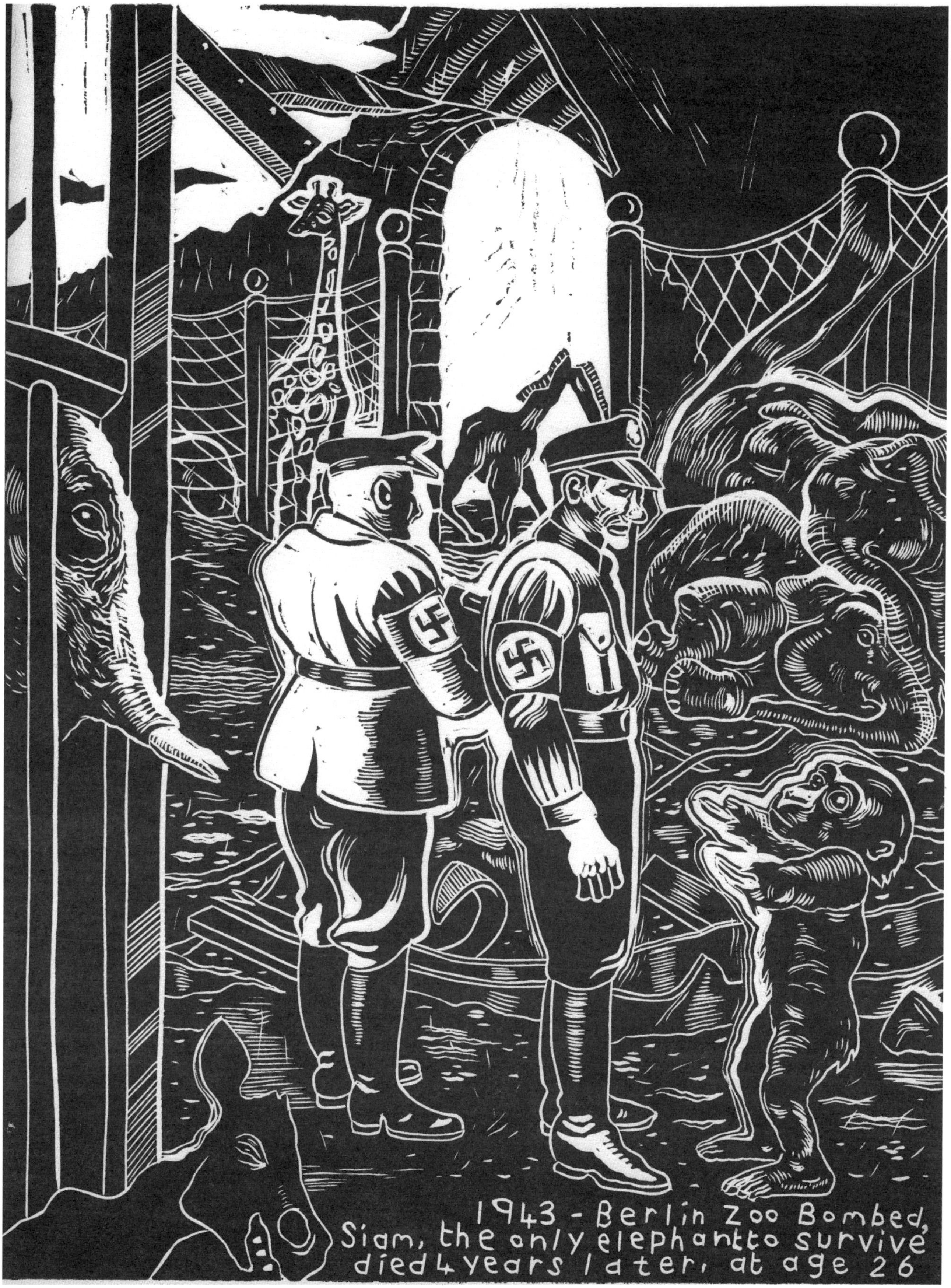
1943 - Berlin Zoo Bombed,
Siam, the only elephant to survive
died 4 years later, at age 26

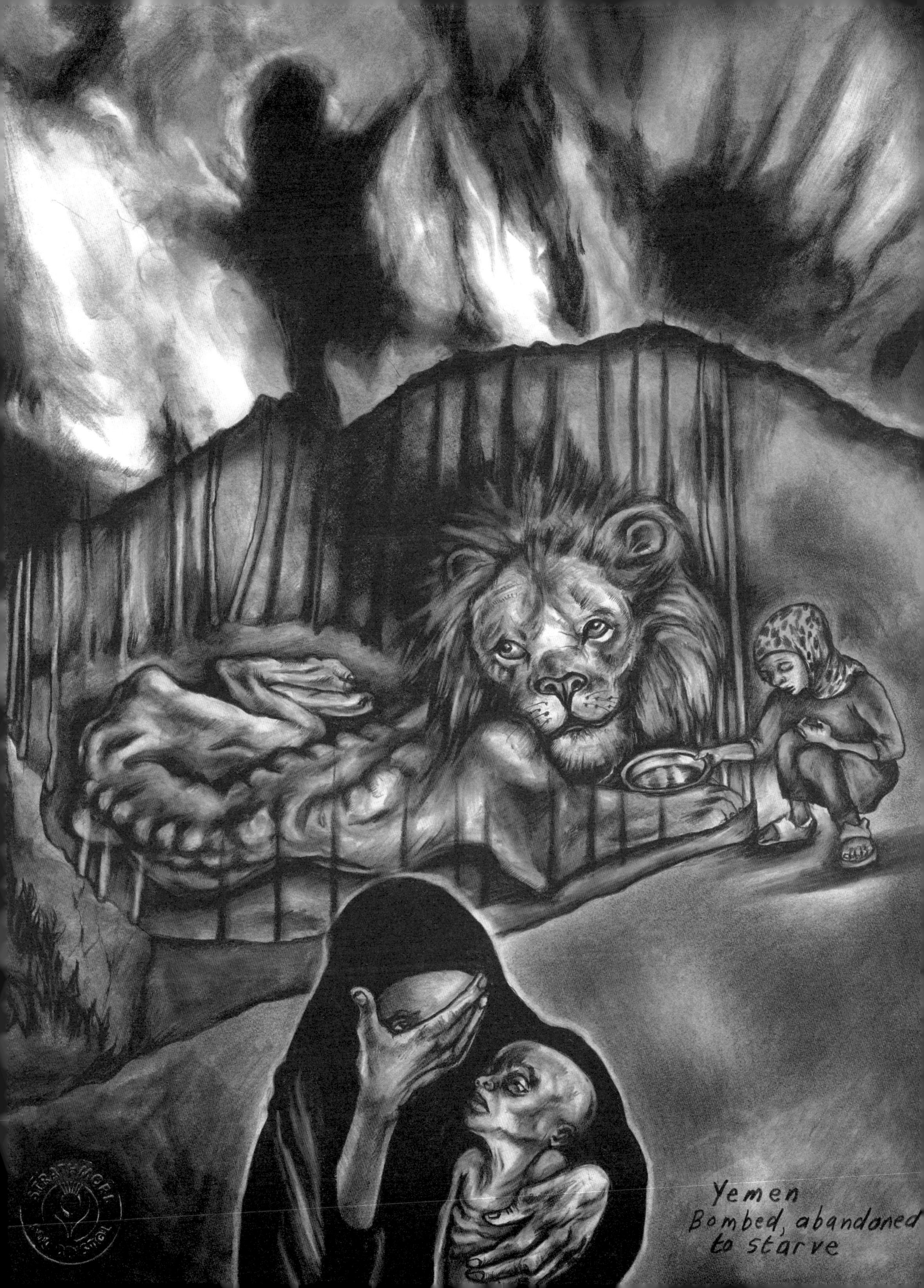
Yemen
Bombed, abandaned to starve

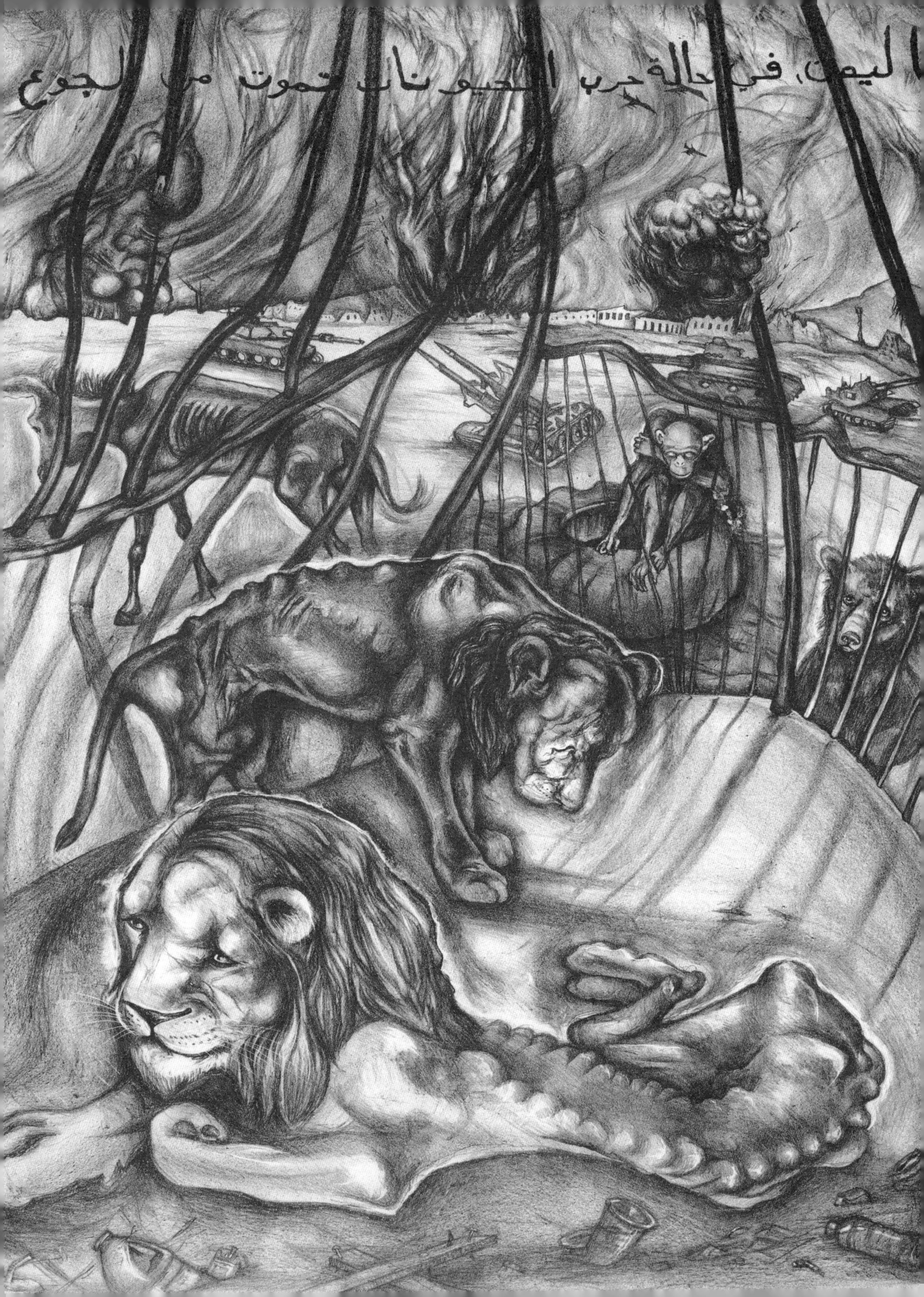
اليمن، في حالة حرب الحيوانات تموت من الجوع

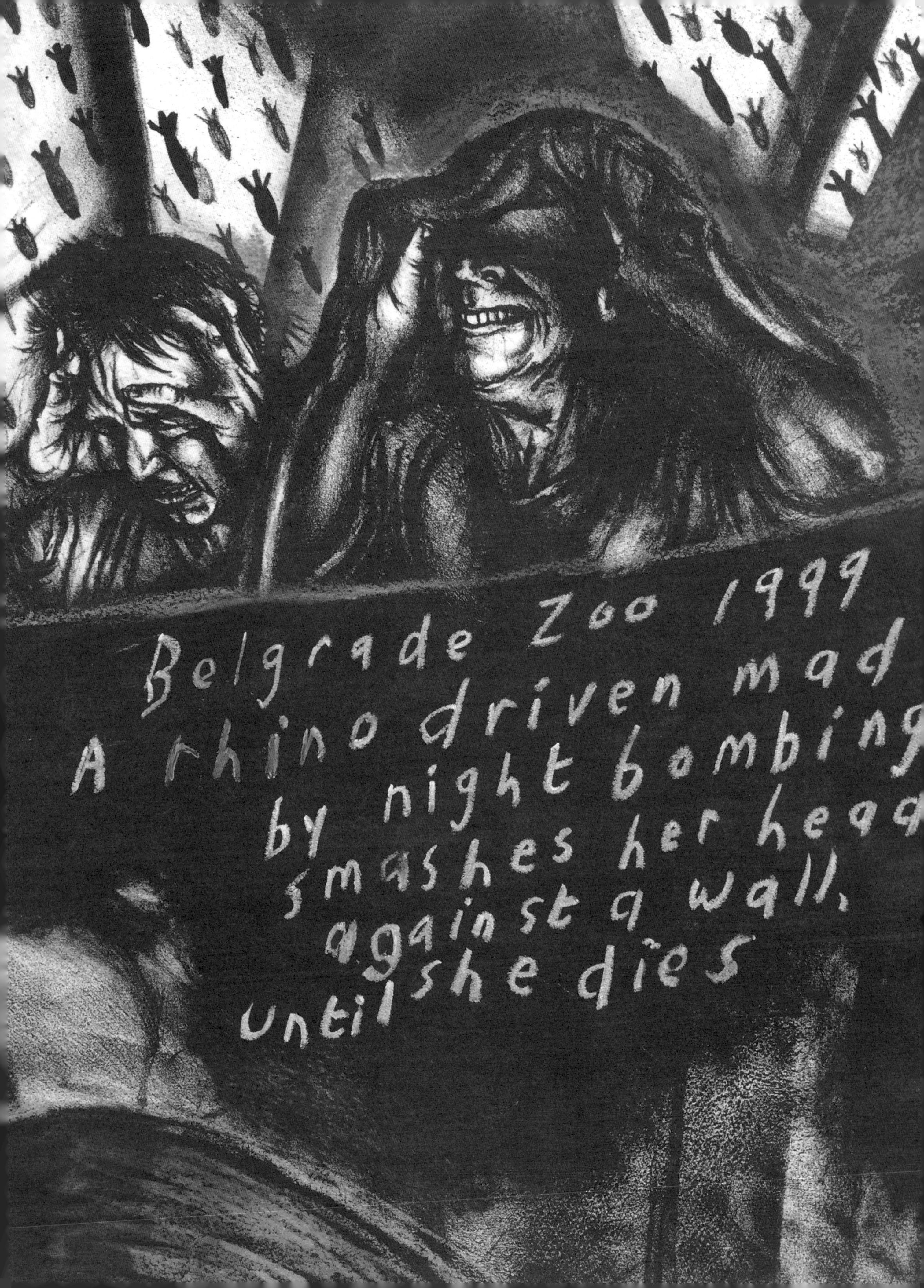
Belgrade Zoo 1999
A rhino driven mad
by night bombing
smashes her head
against a wall,
until she dies

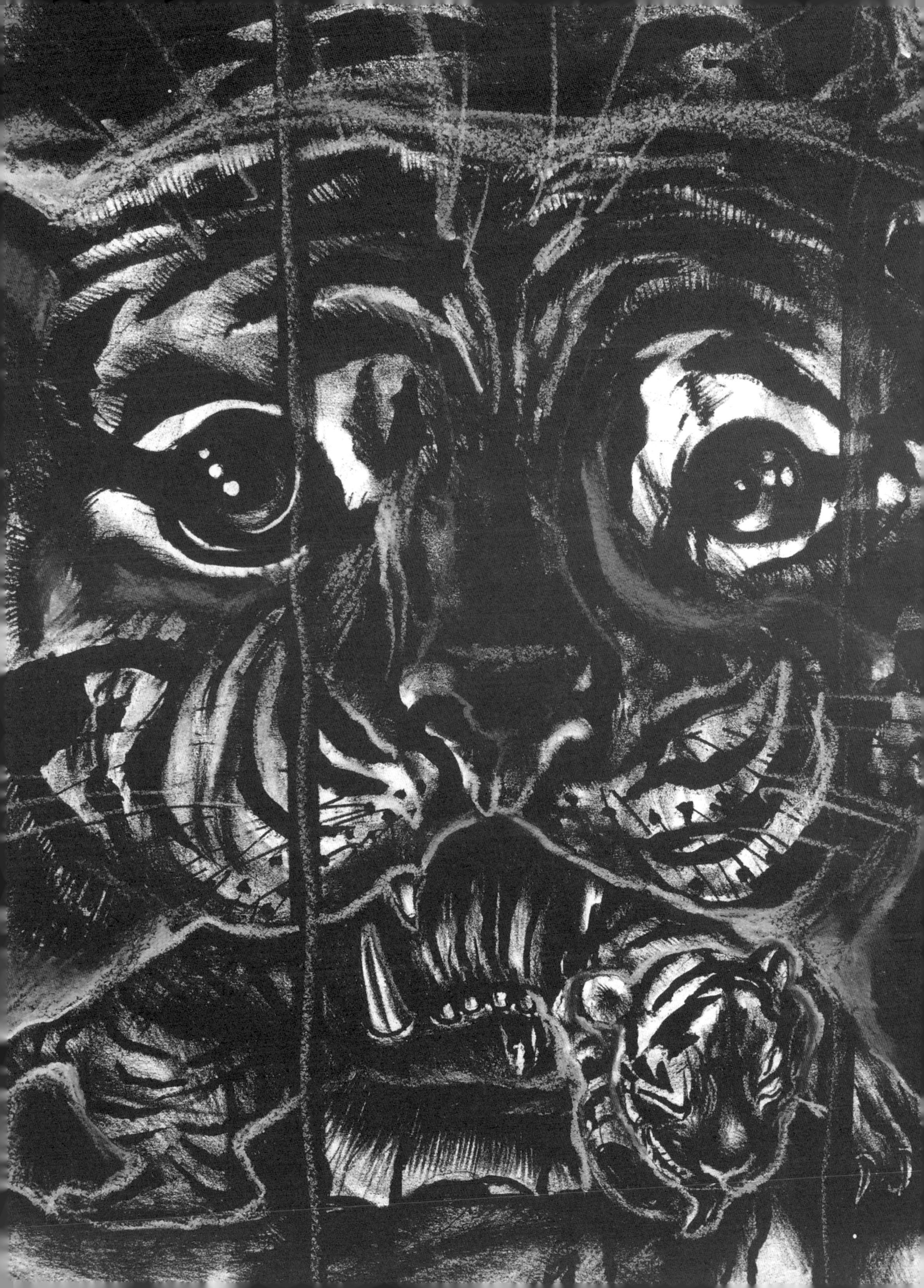

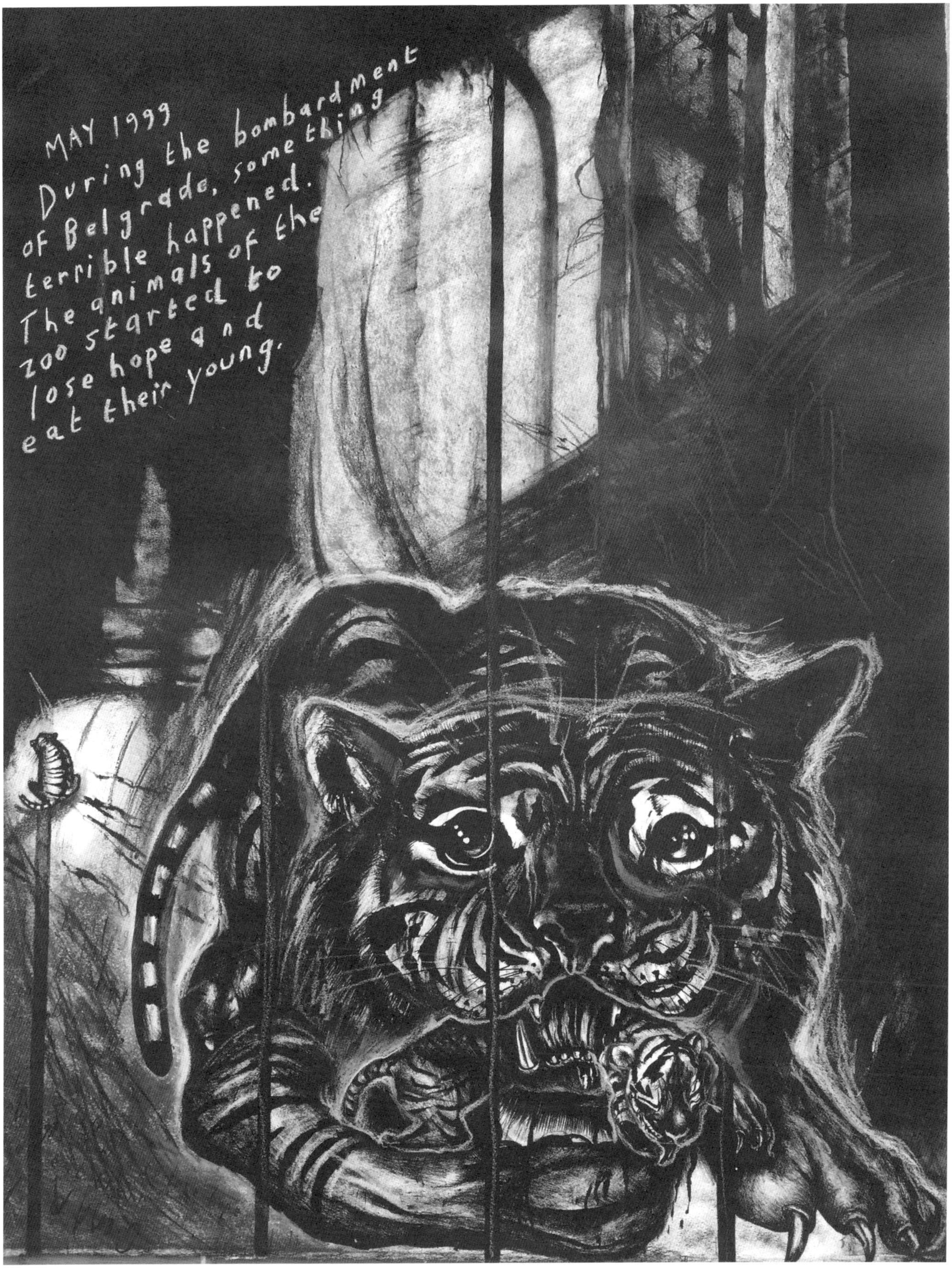
MAY 1999
During the bombardment of Belgrade, something terrible happened. The animals of the zoo started to lose hope and eat their young.

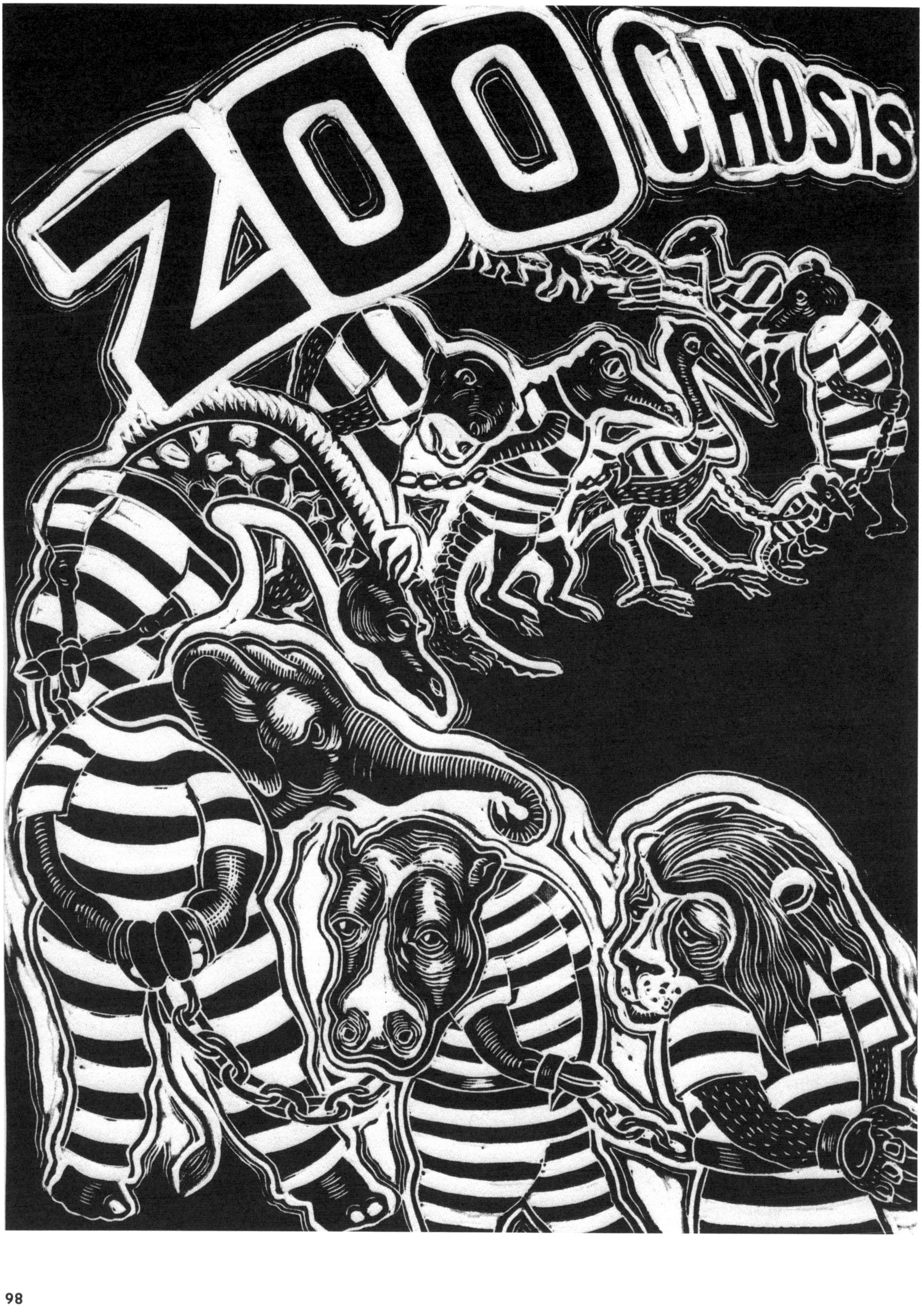
ZOOCHOSIS

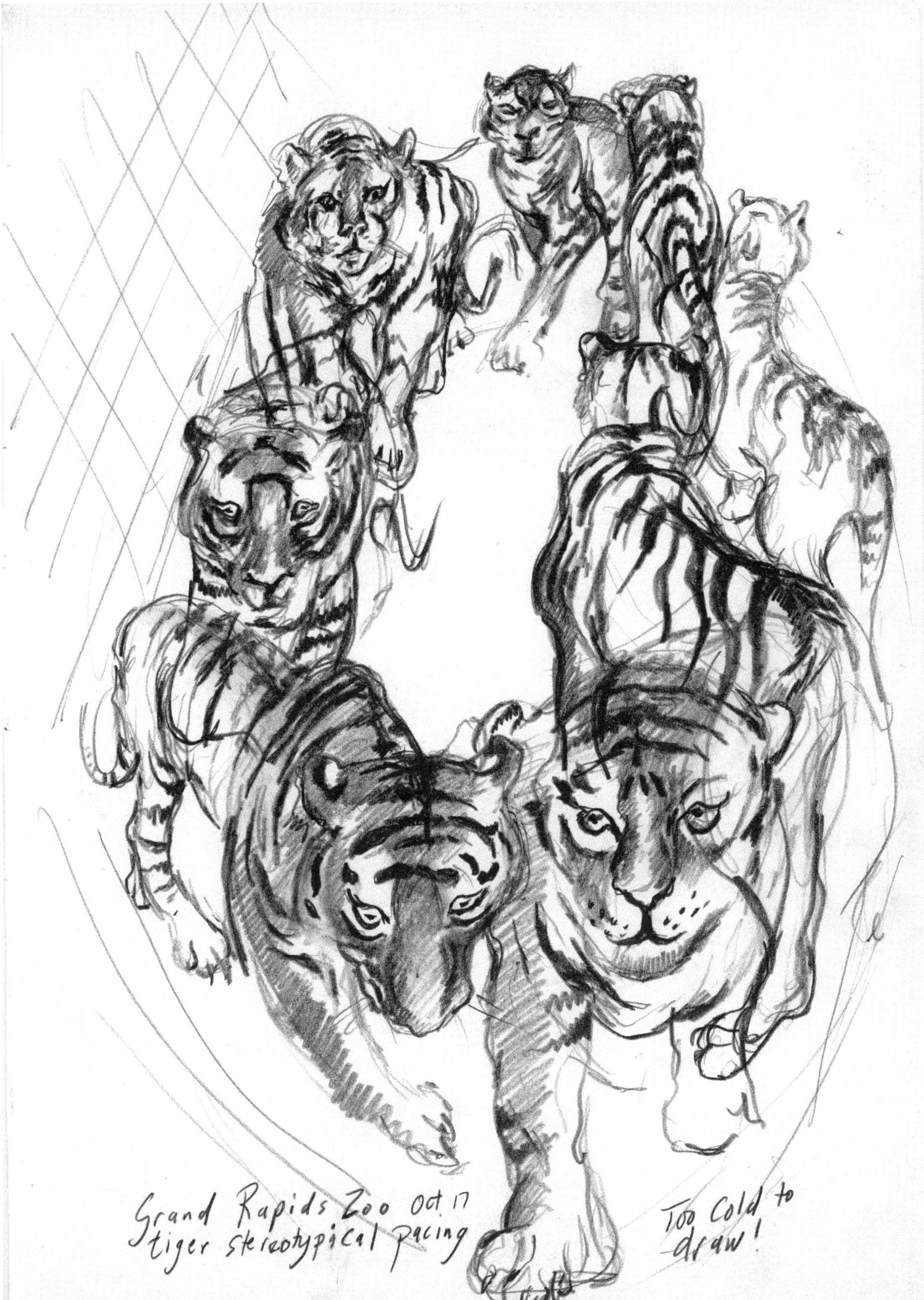
Grand Rapids Zoo Oct 17
tiger stereotypical pacing
Too cold to draw!

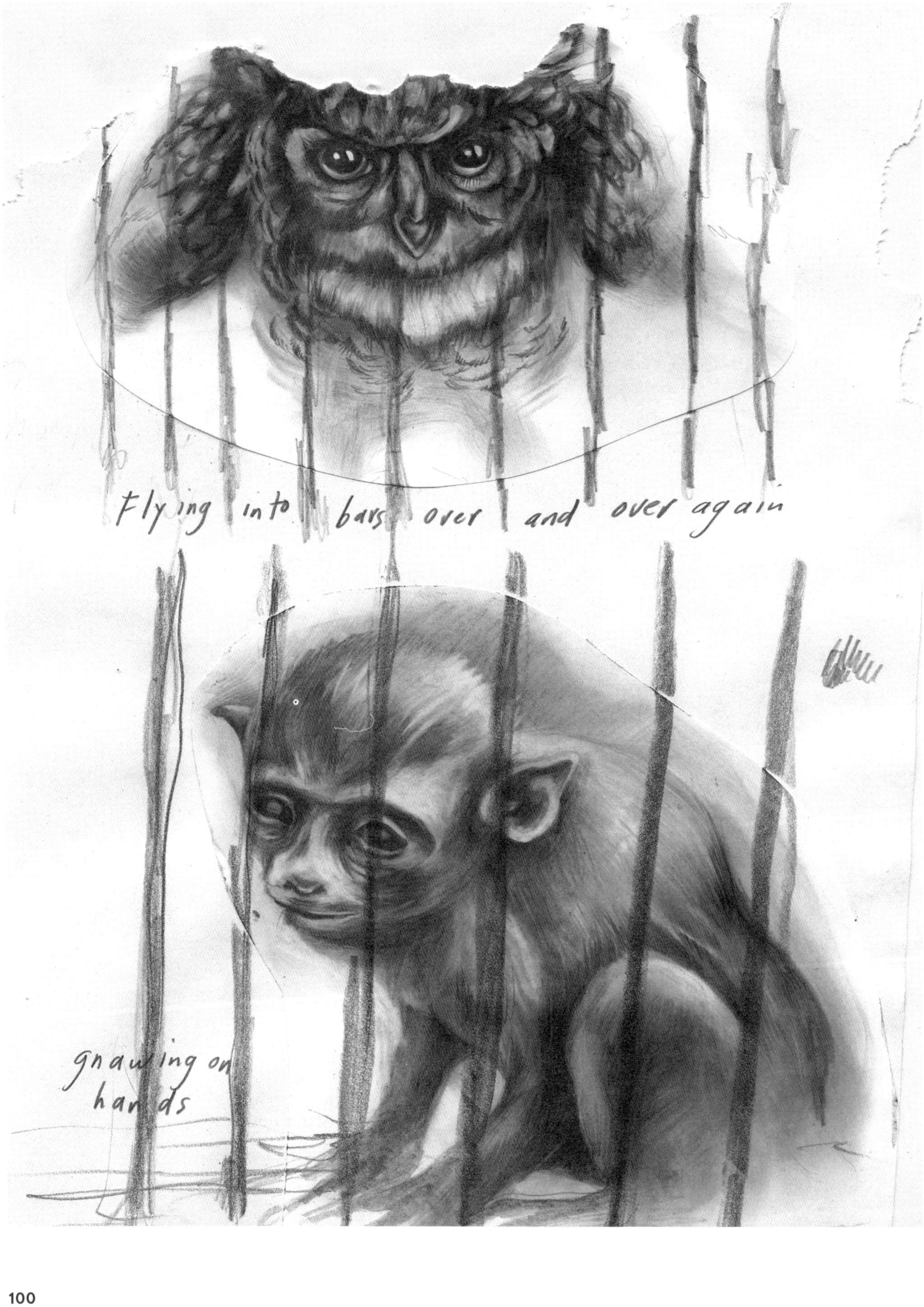
Flying into bars over and over again
gnawing on hands

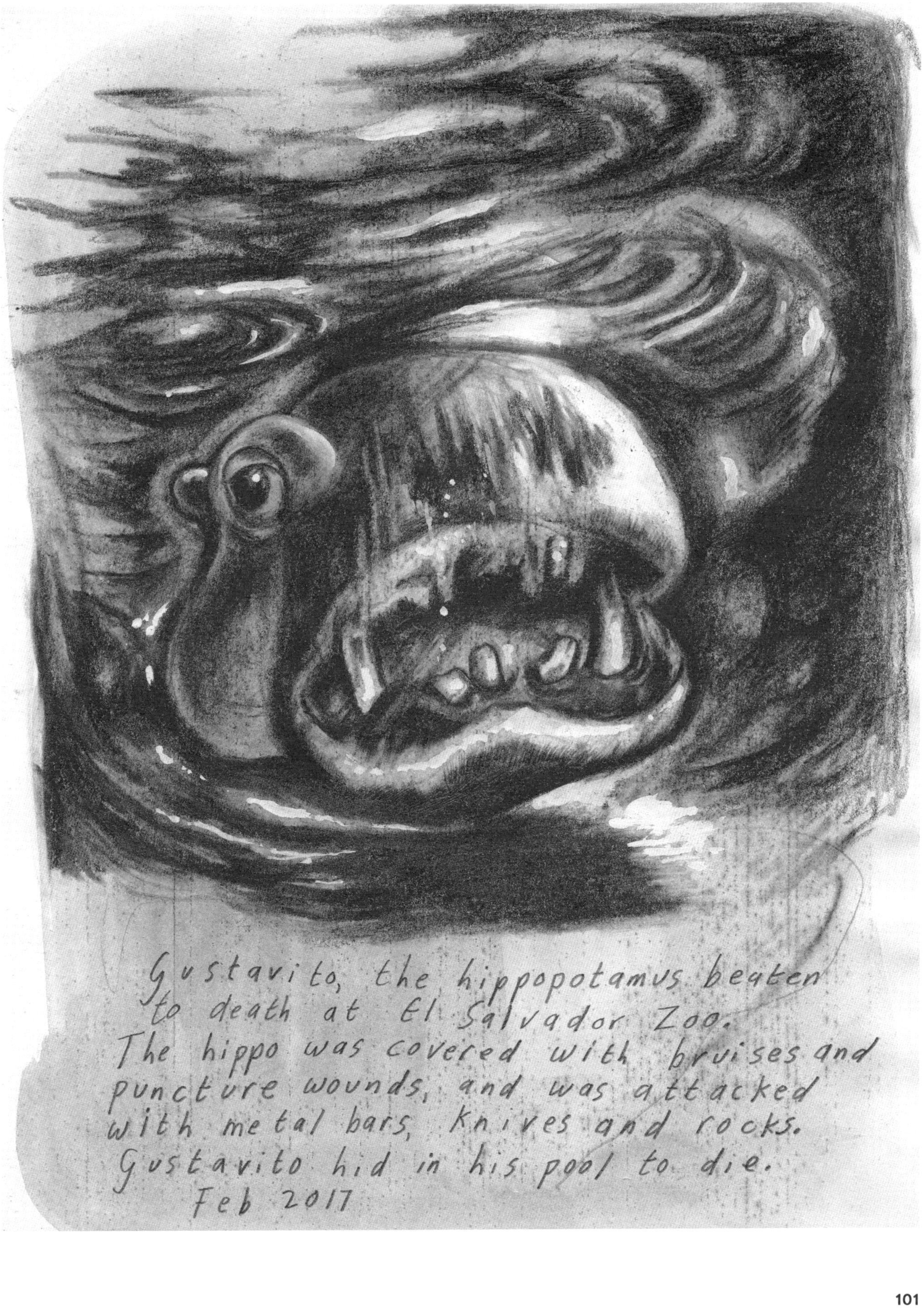
Gustavito, the hippopotamus beaten to death at El Salvador Zoo.
The hippo was covered with bruises and puncture wounds, and was attacked with metal bars, knives and rocks.
Gustavito hid in his pool to die.
Feb 2017

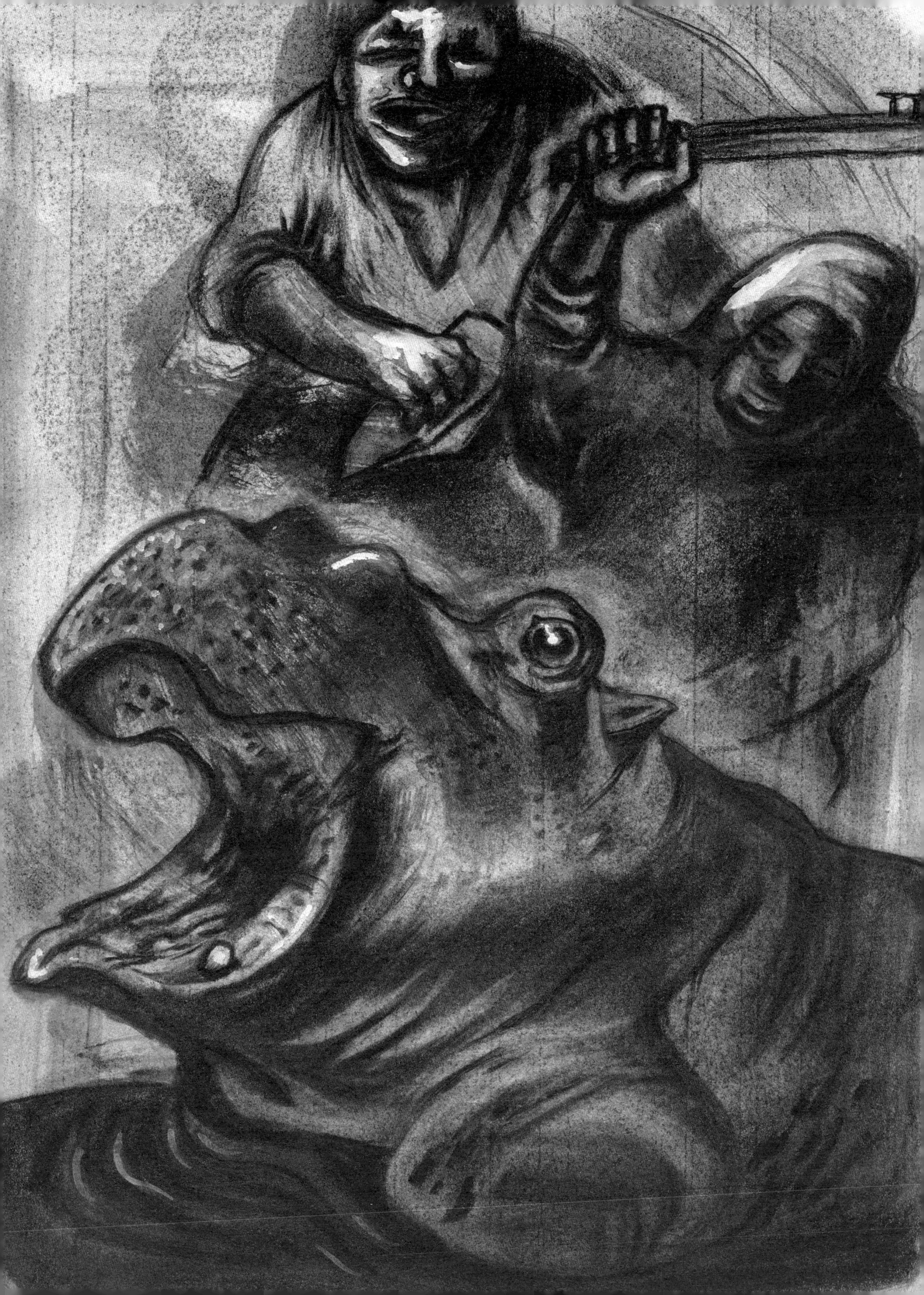

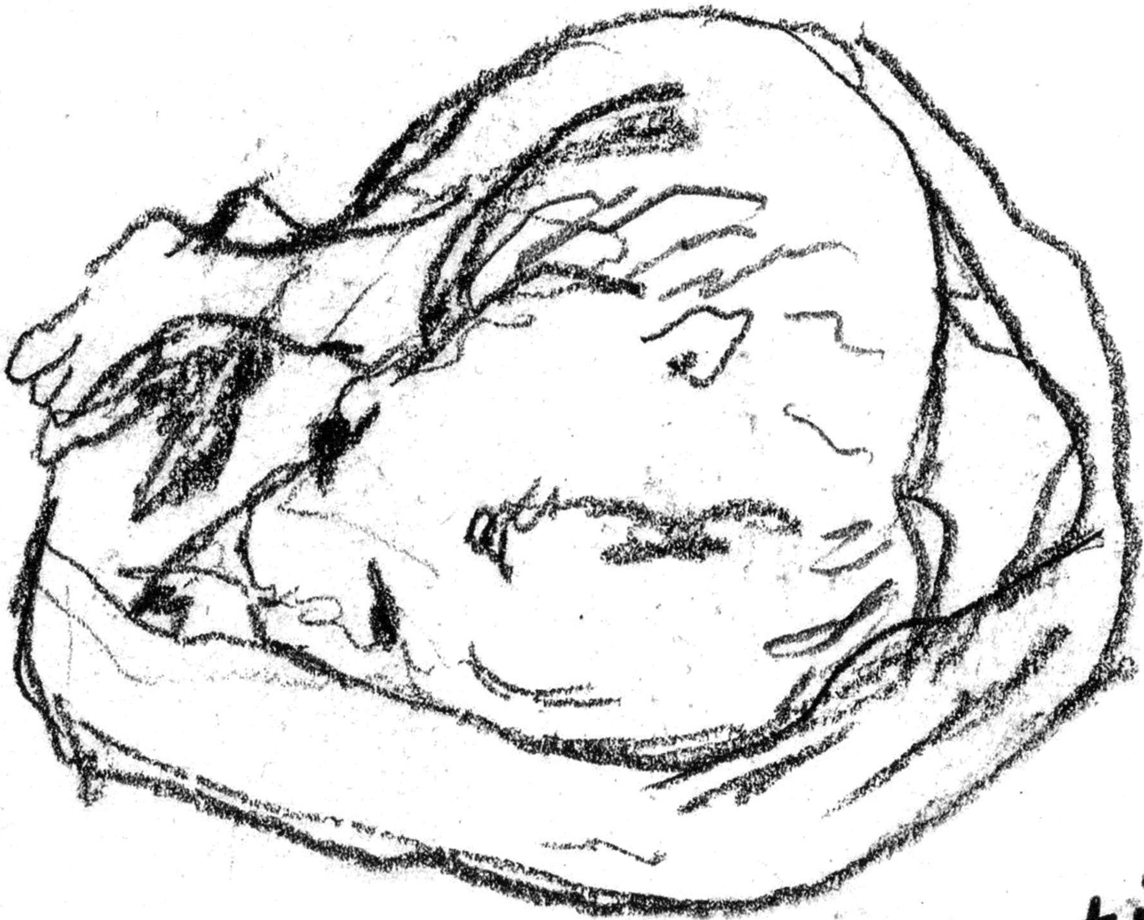

Lion sleeping in plastic pool

Kid screaming "Wake Up"

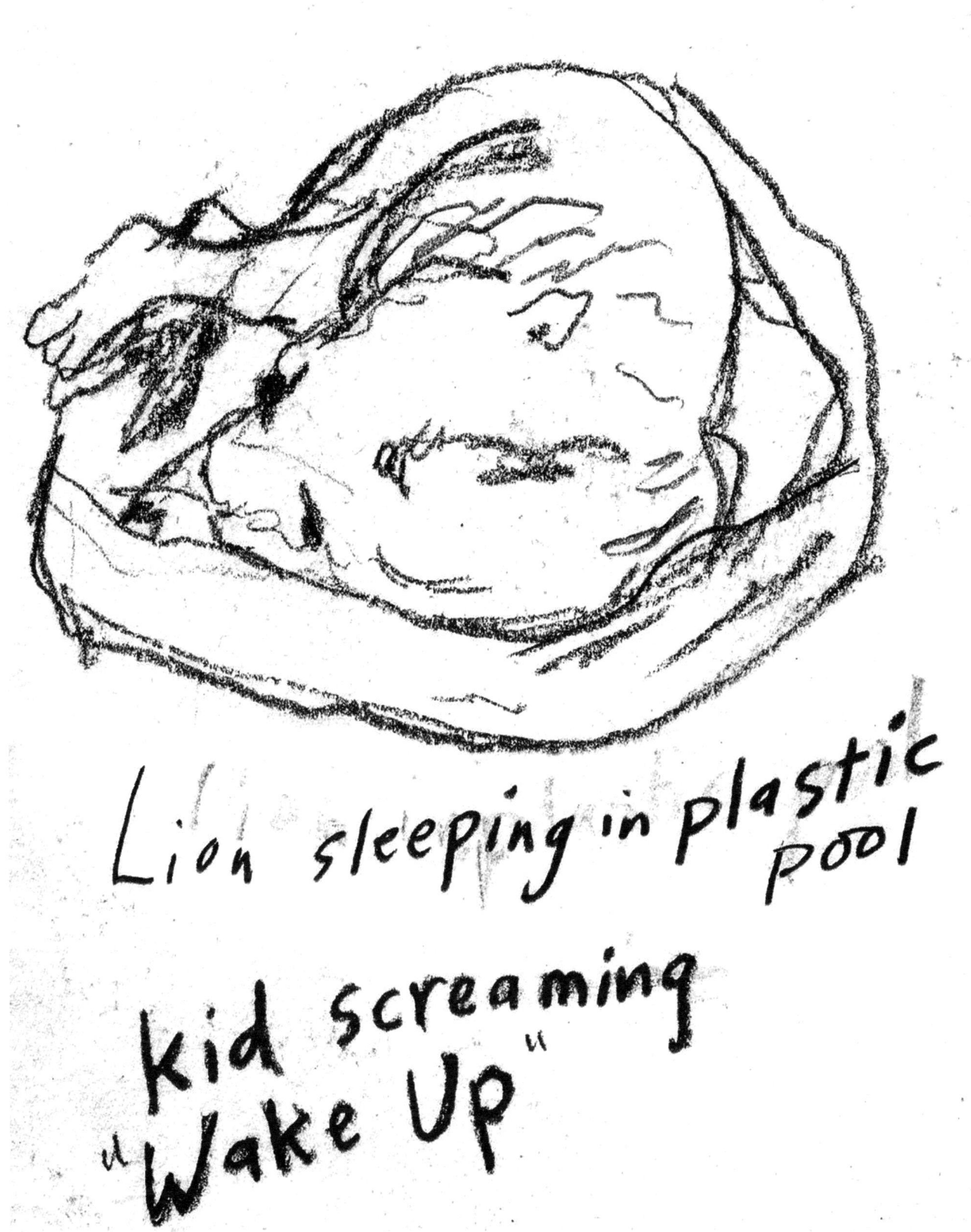

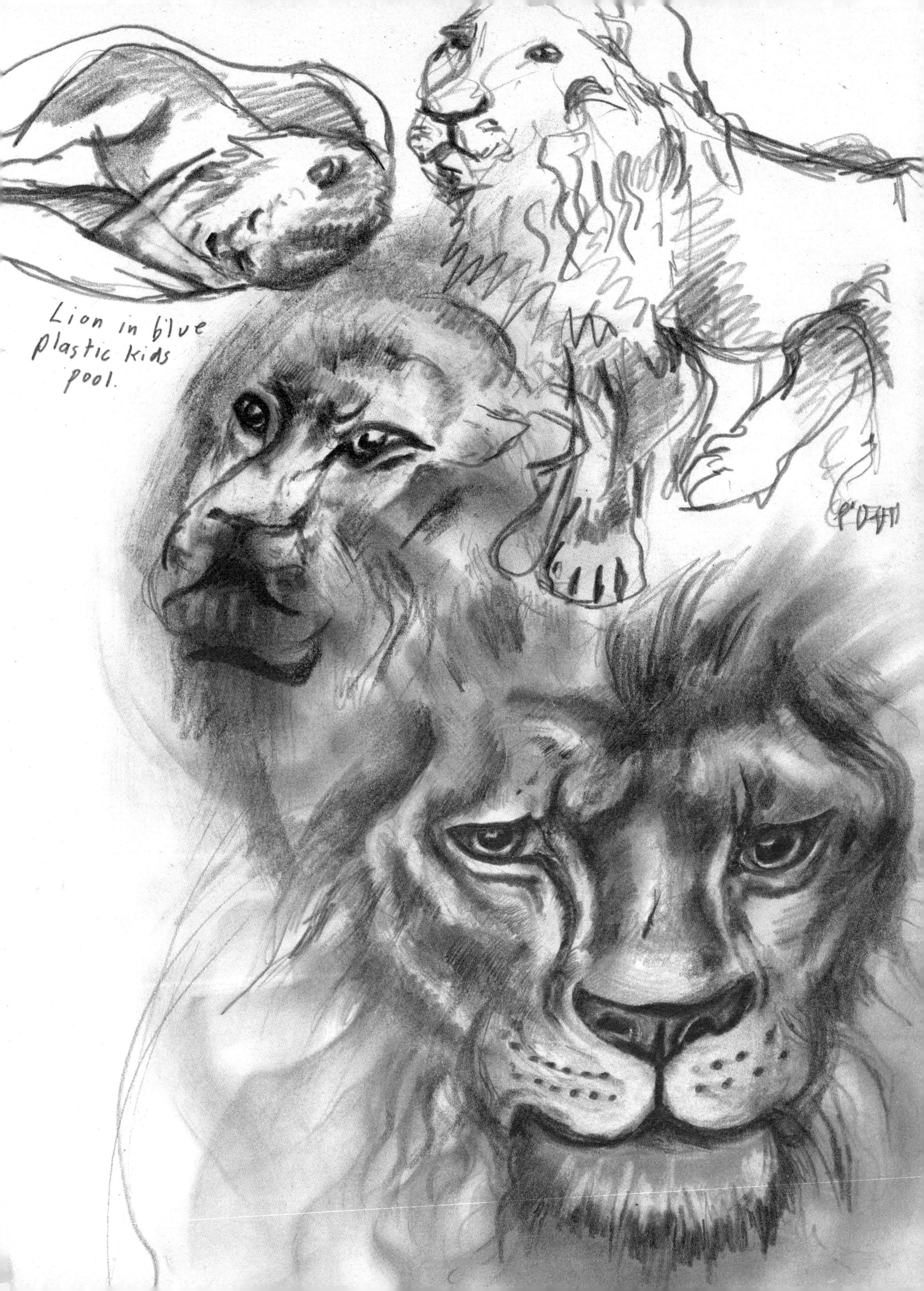
Lion in blue plastic kids pool.

Feb 2018 Fuzhou Zoo China. Kangaroo stoned to death for not hopping. Her foot was almost severed. She died after two days of internal bleeding.

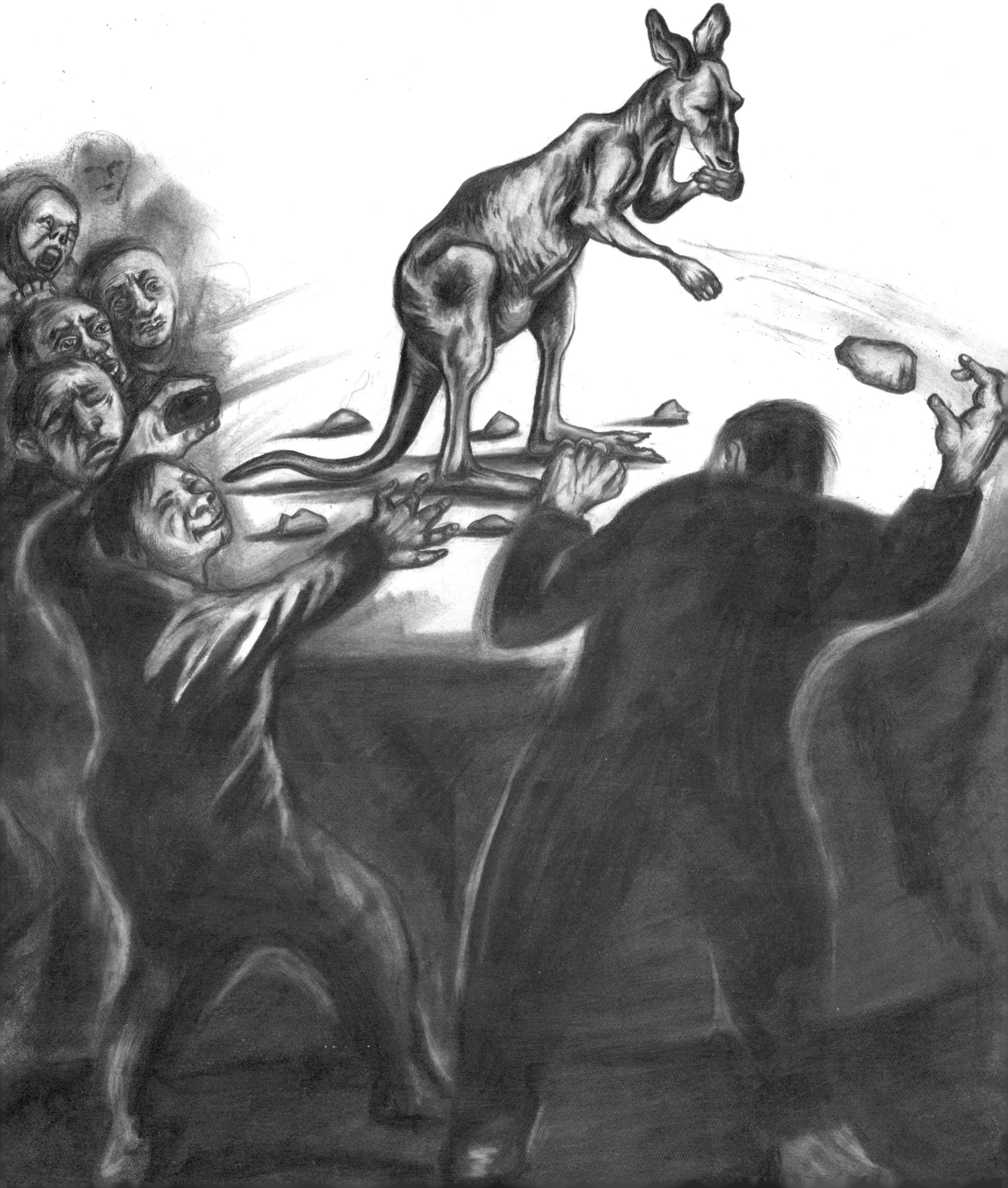

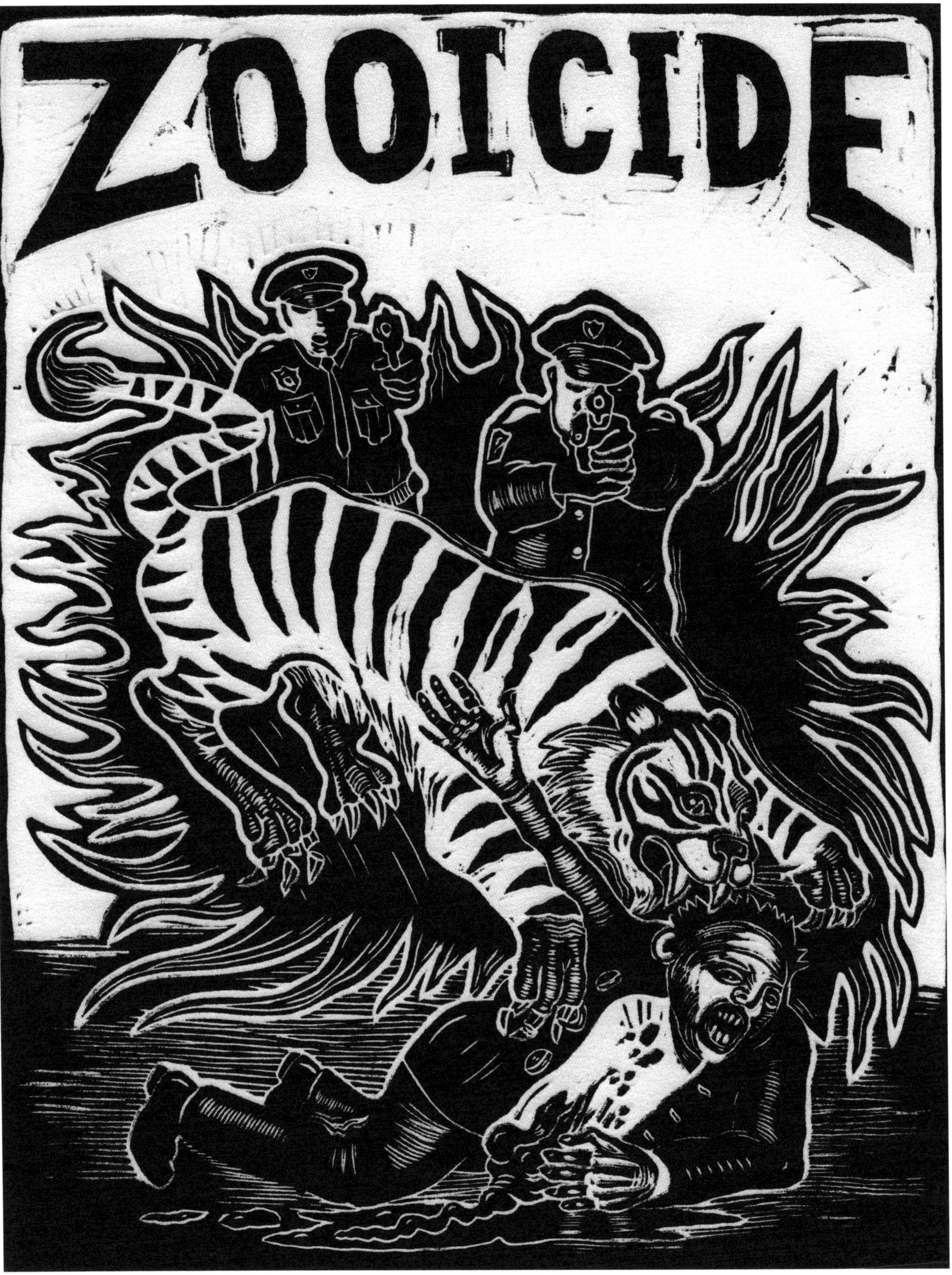
ZOOICIDE

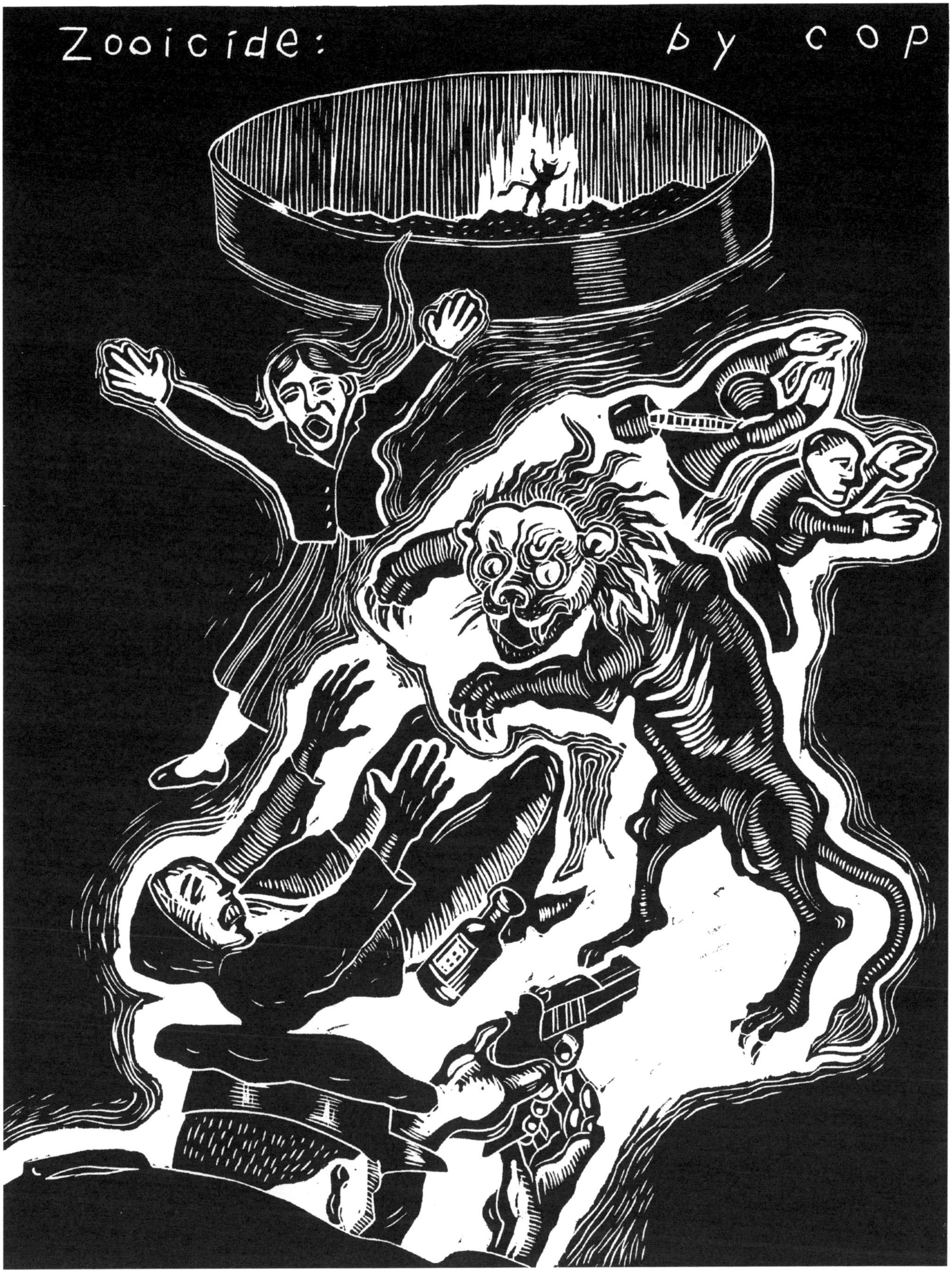
Zooicide: by cop

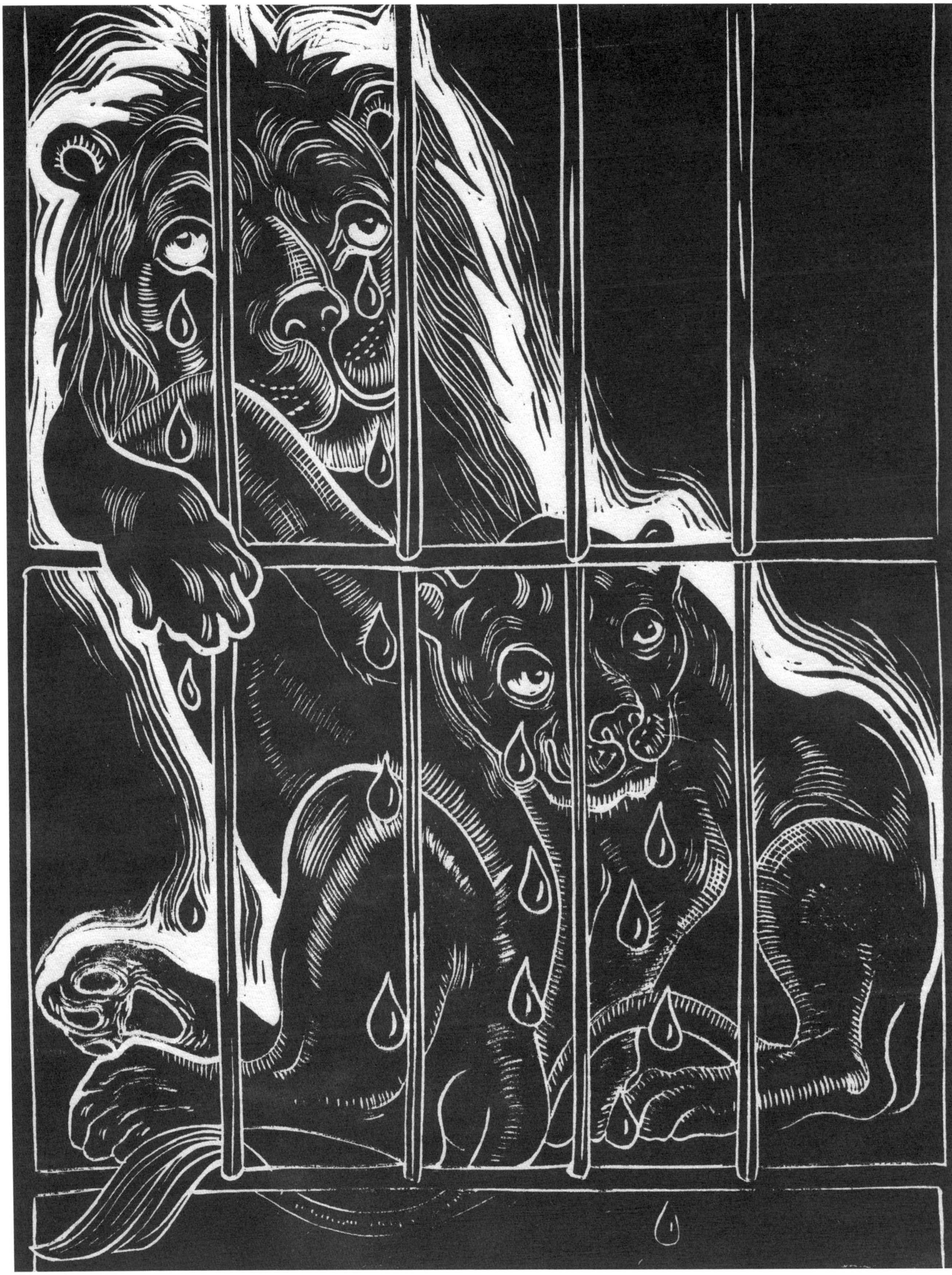

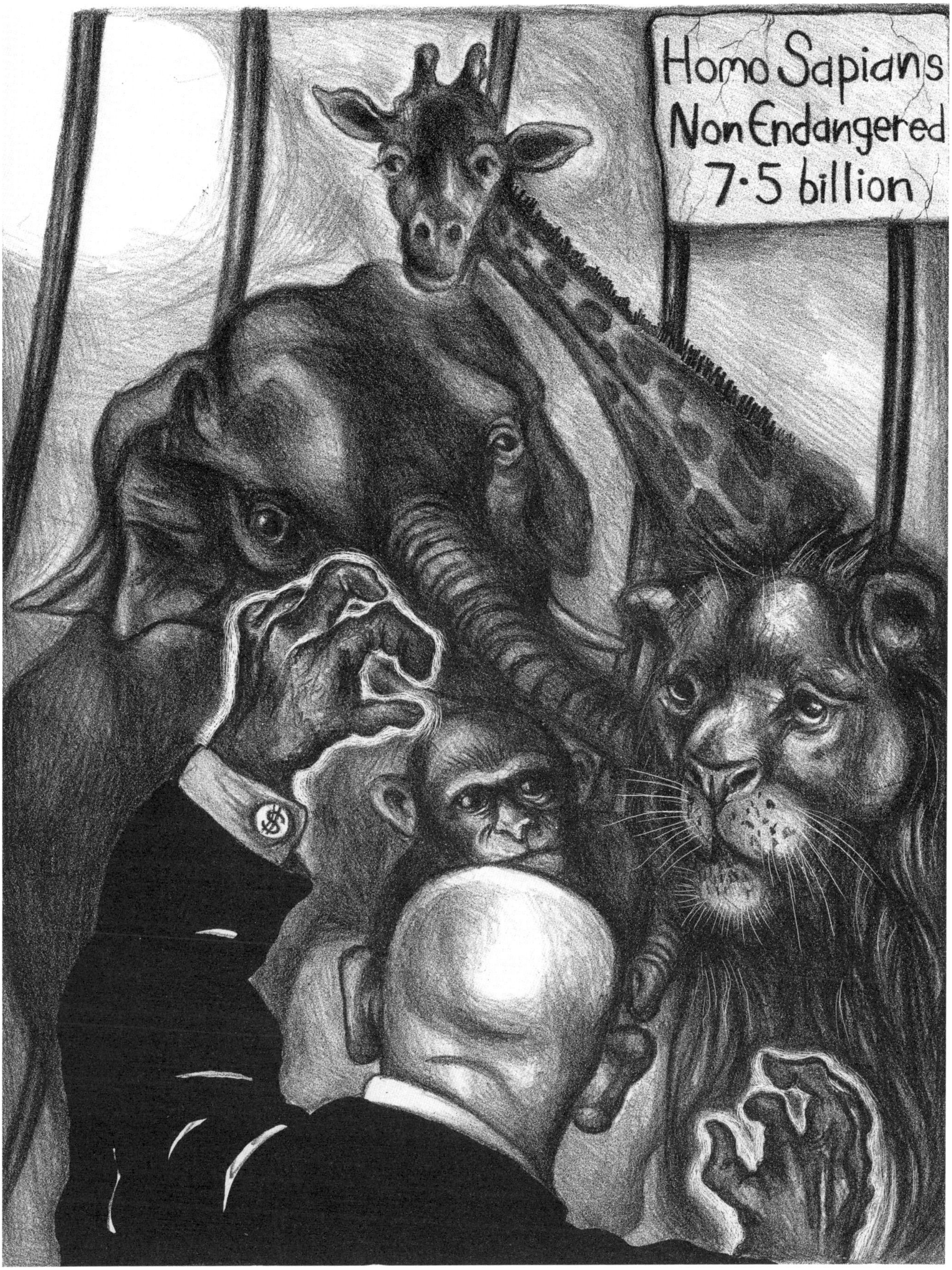
Homo Sapians
Non Endangered
7·5 billion

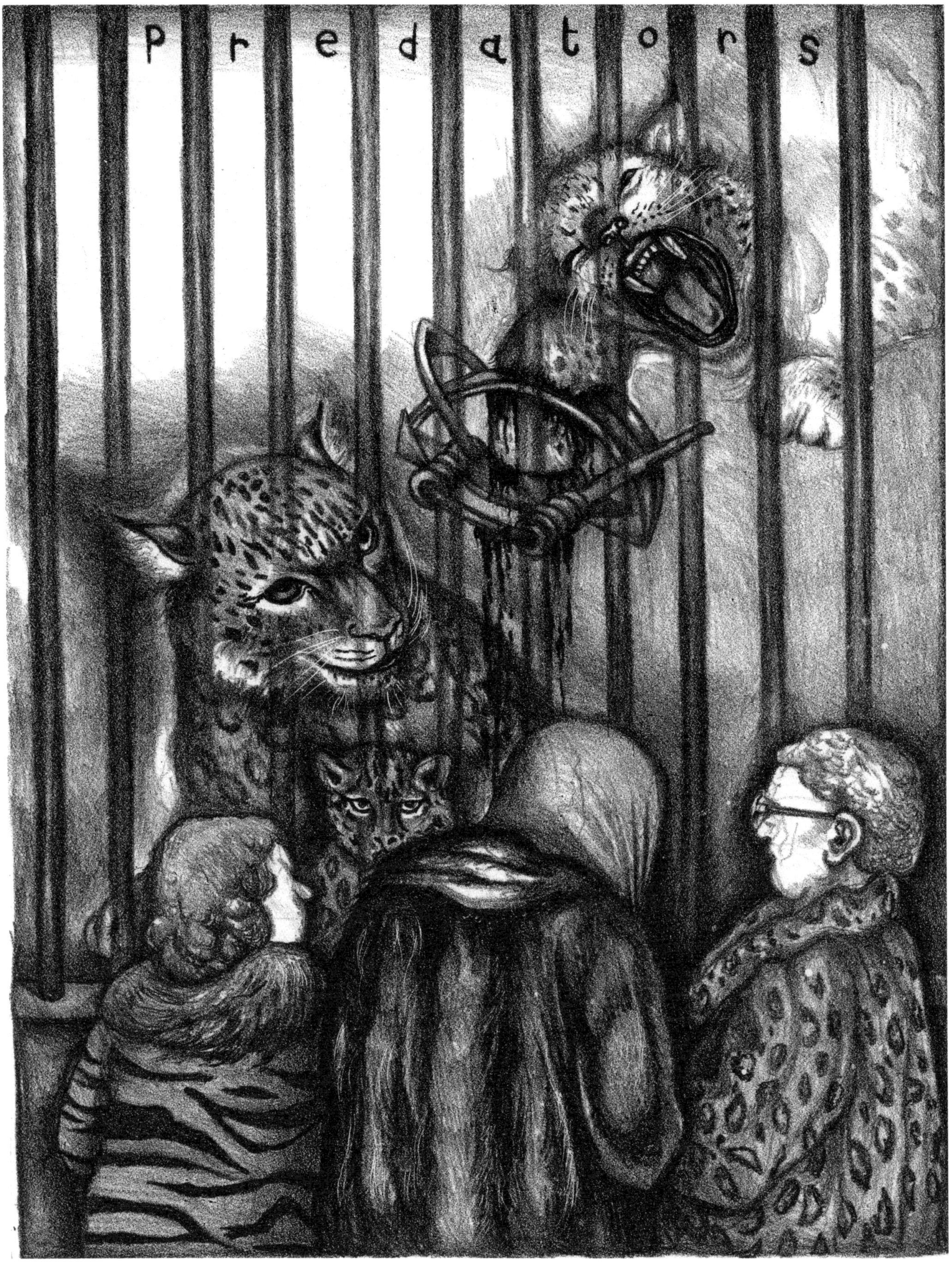
predators

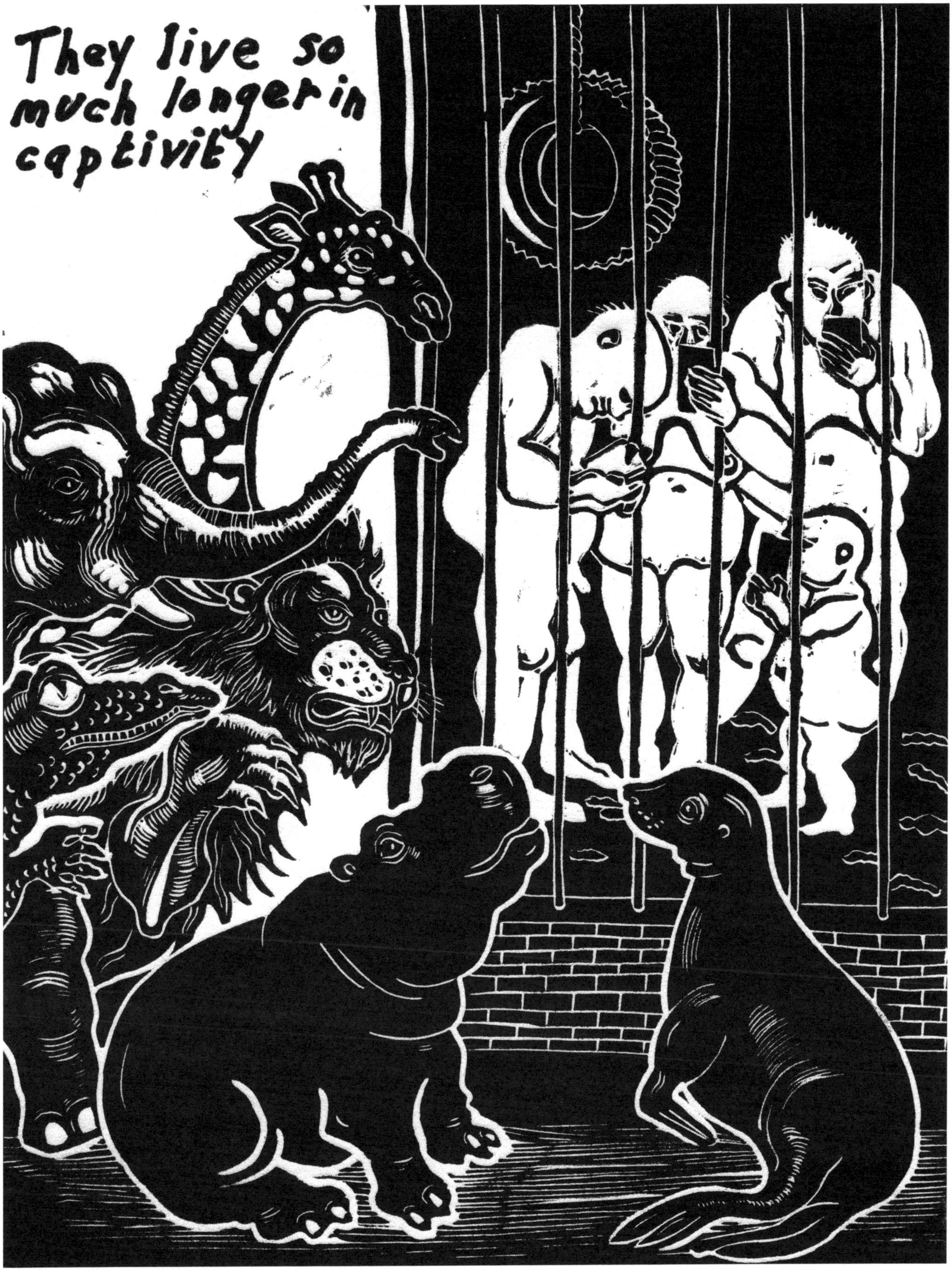
They live so much longer in captivity

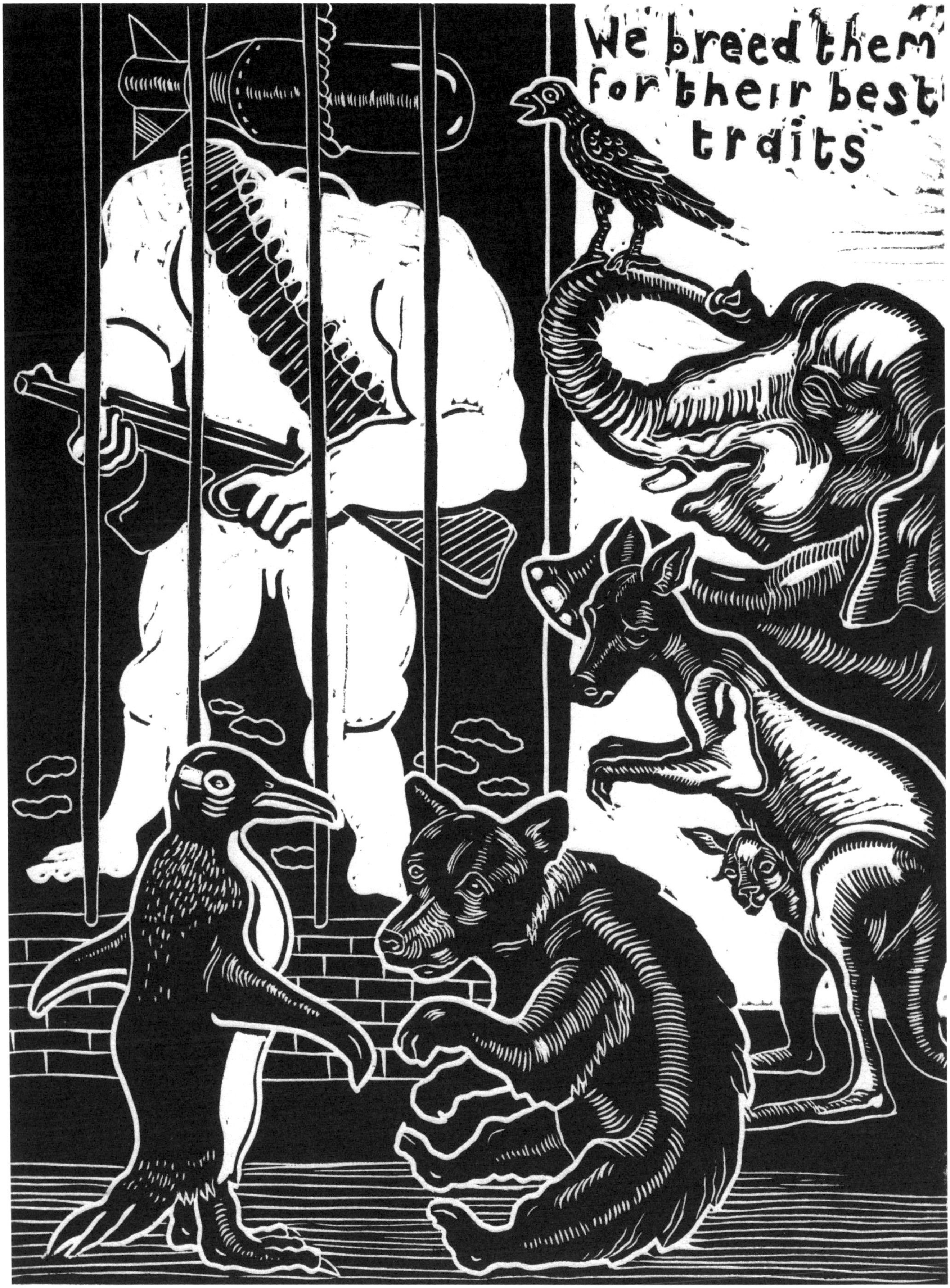
We breed them for their best traits

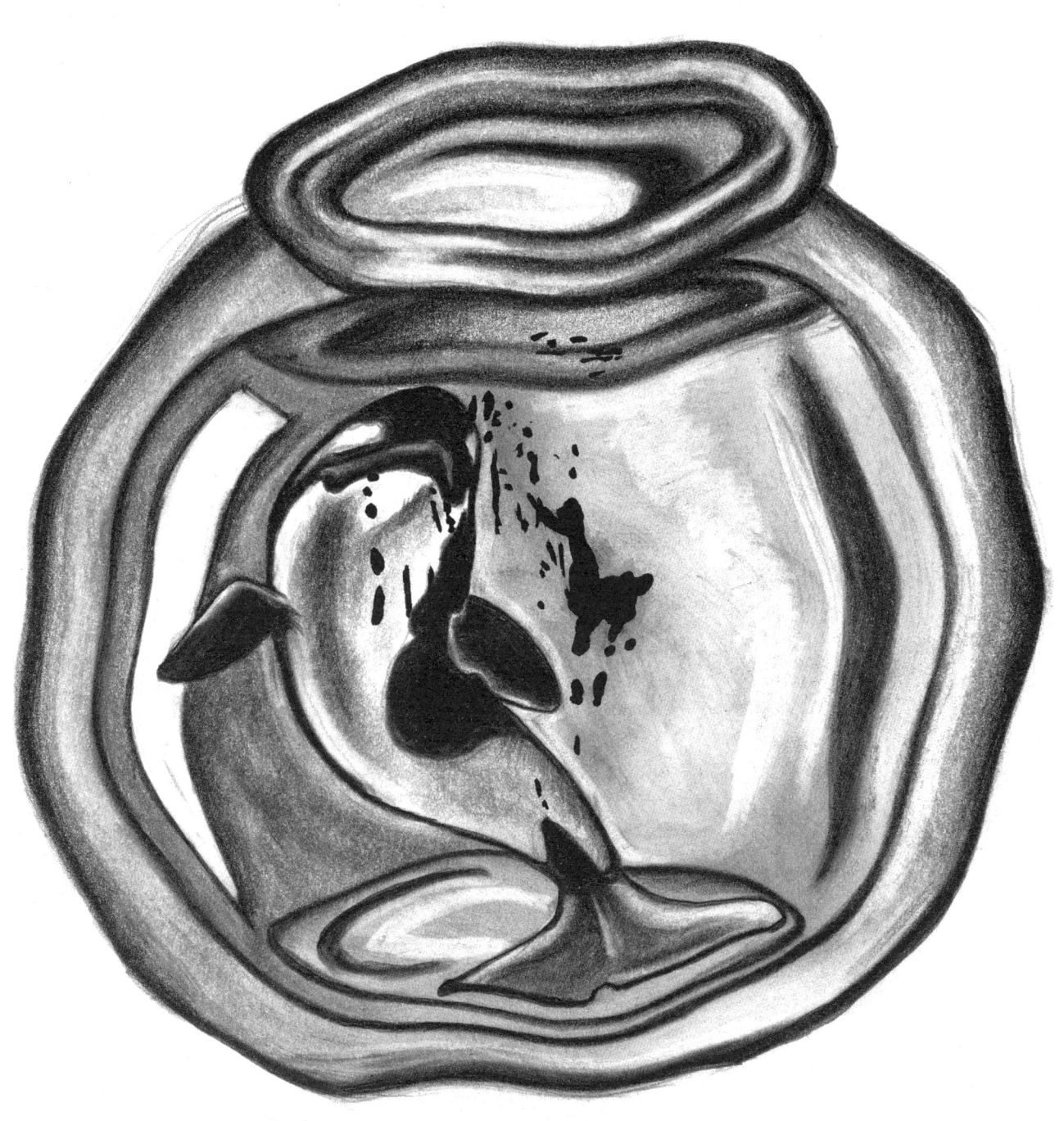

Zooicide: Miami Seaquarium 1980
Hugo the Orca killed himself by repeatedly smashing his head into the walls of his tank.

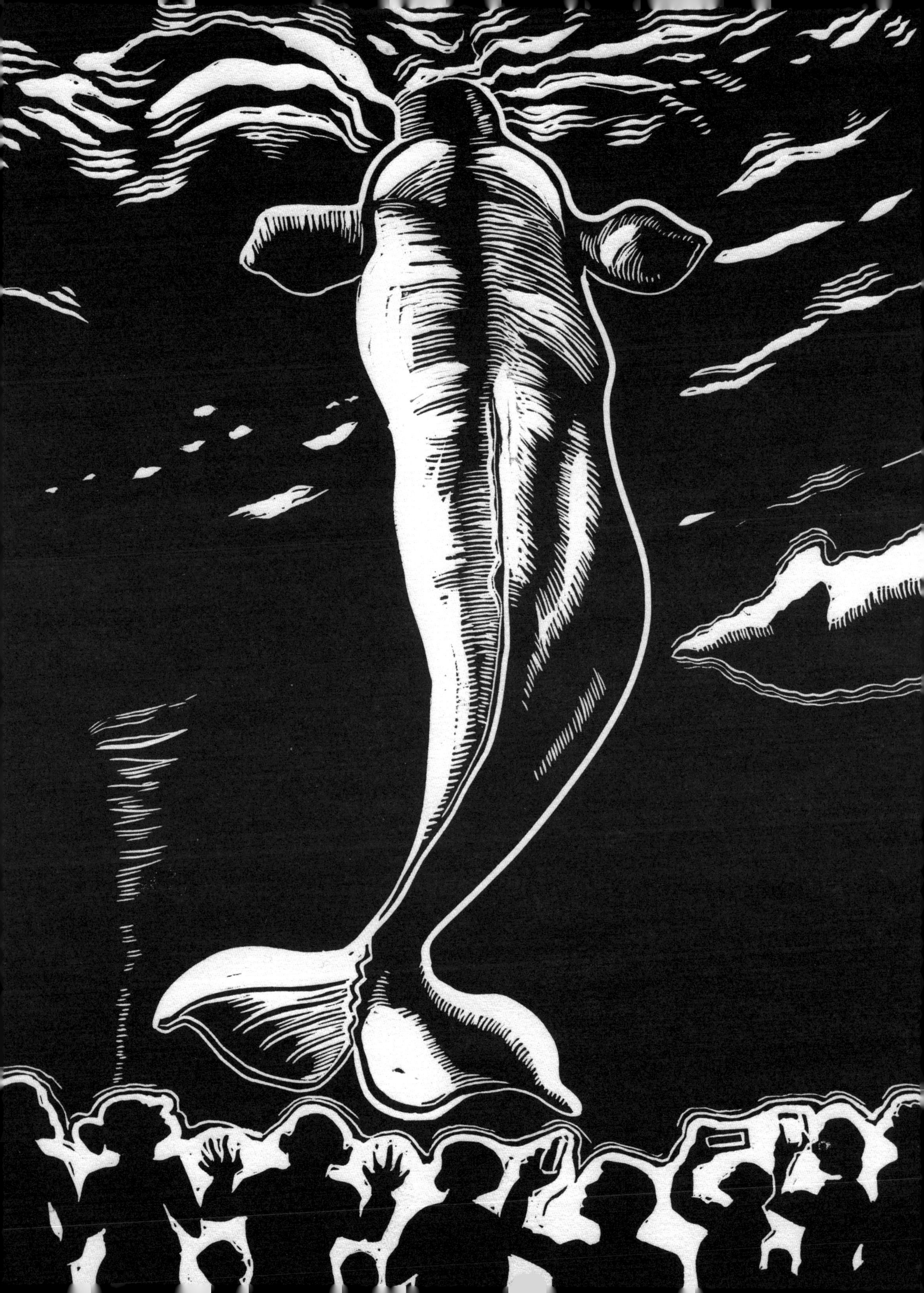

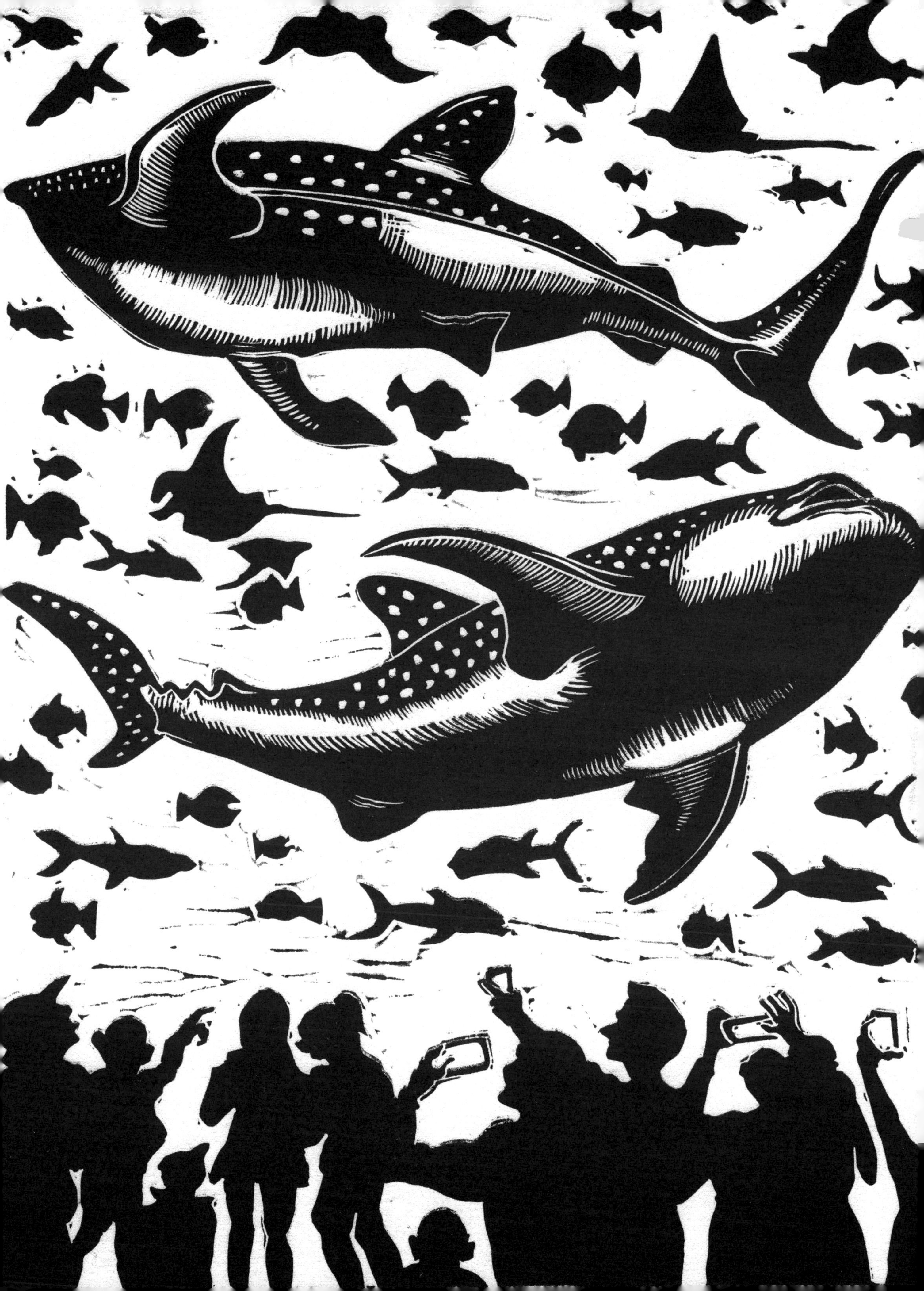

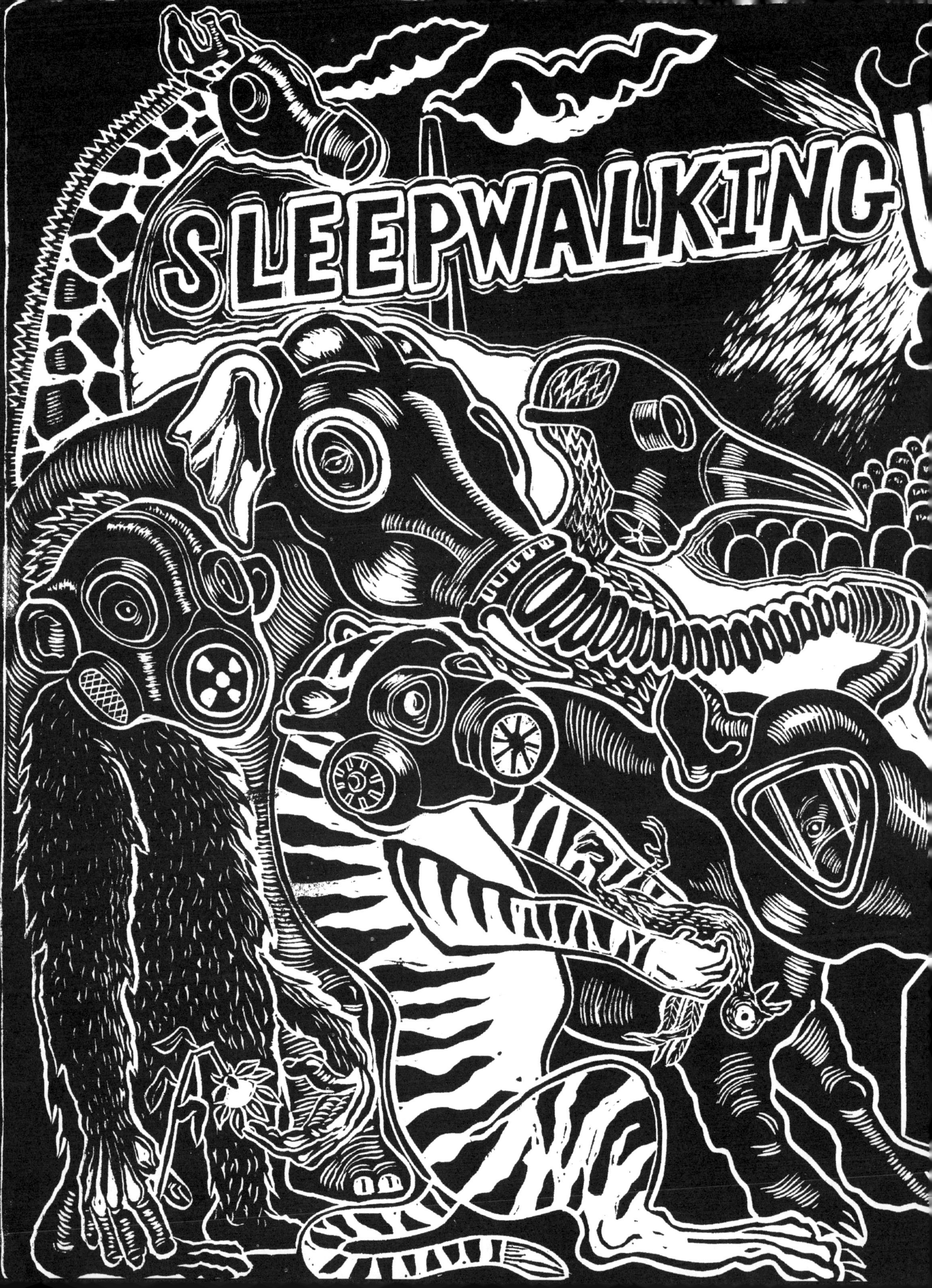
SLEEPWALKING

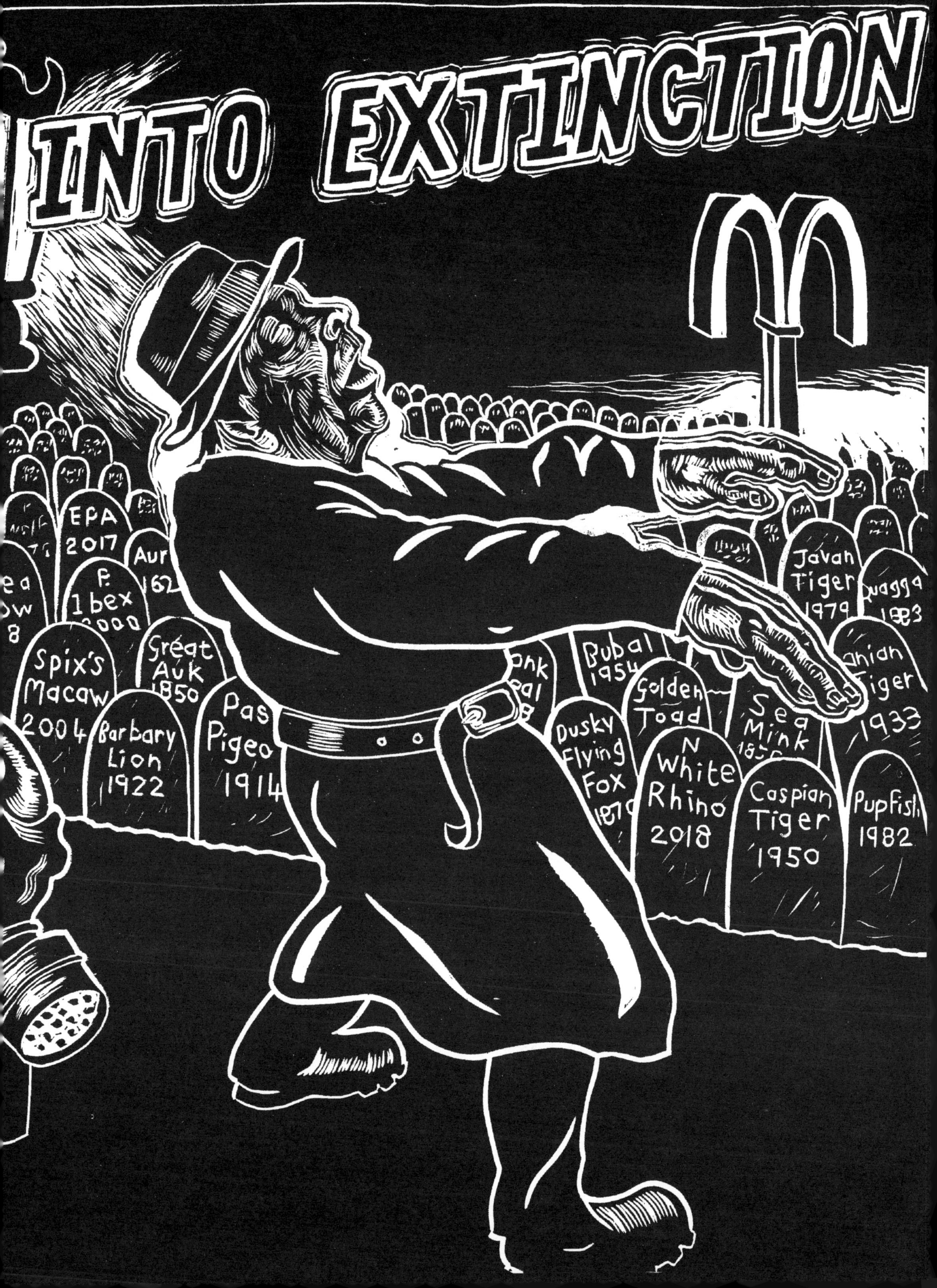
INTO EXTINCTION
EPA 2017
Spix's Macaw 2004
Great Auk 1850
Barbary Lion 1922
Bubal 1954
Golden Toad
Dusky Flying Fox
N White Rhino 2018
Caspian Tiger 1950
Pupfish 1982
Javan Tiger 1979
Quagga 1883
Sea Mink

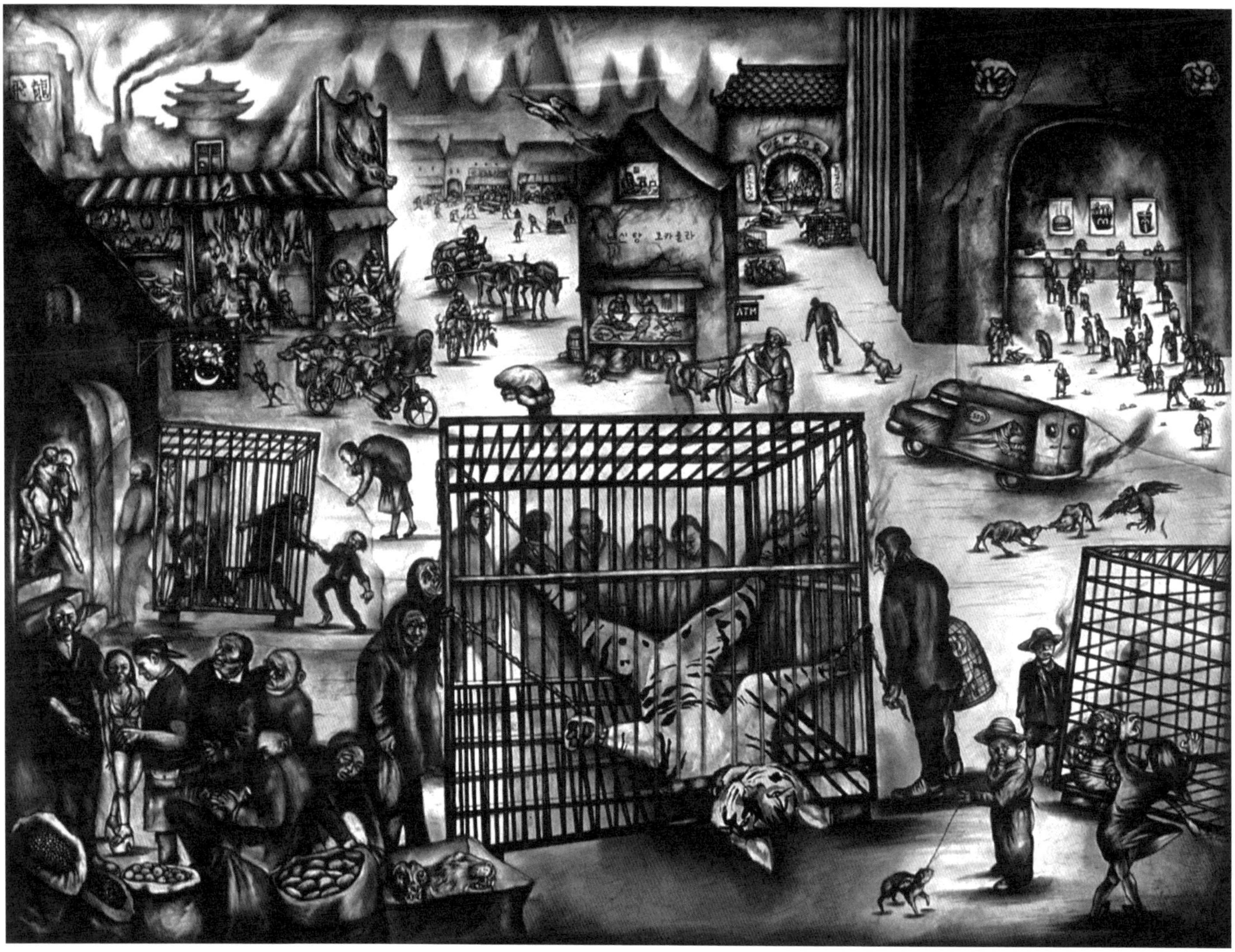
ATM

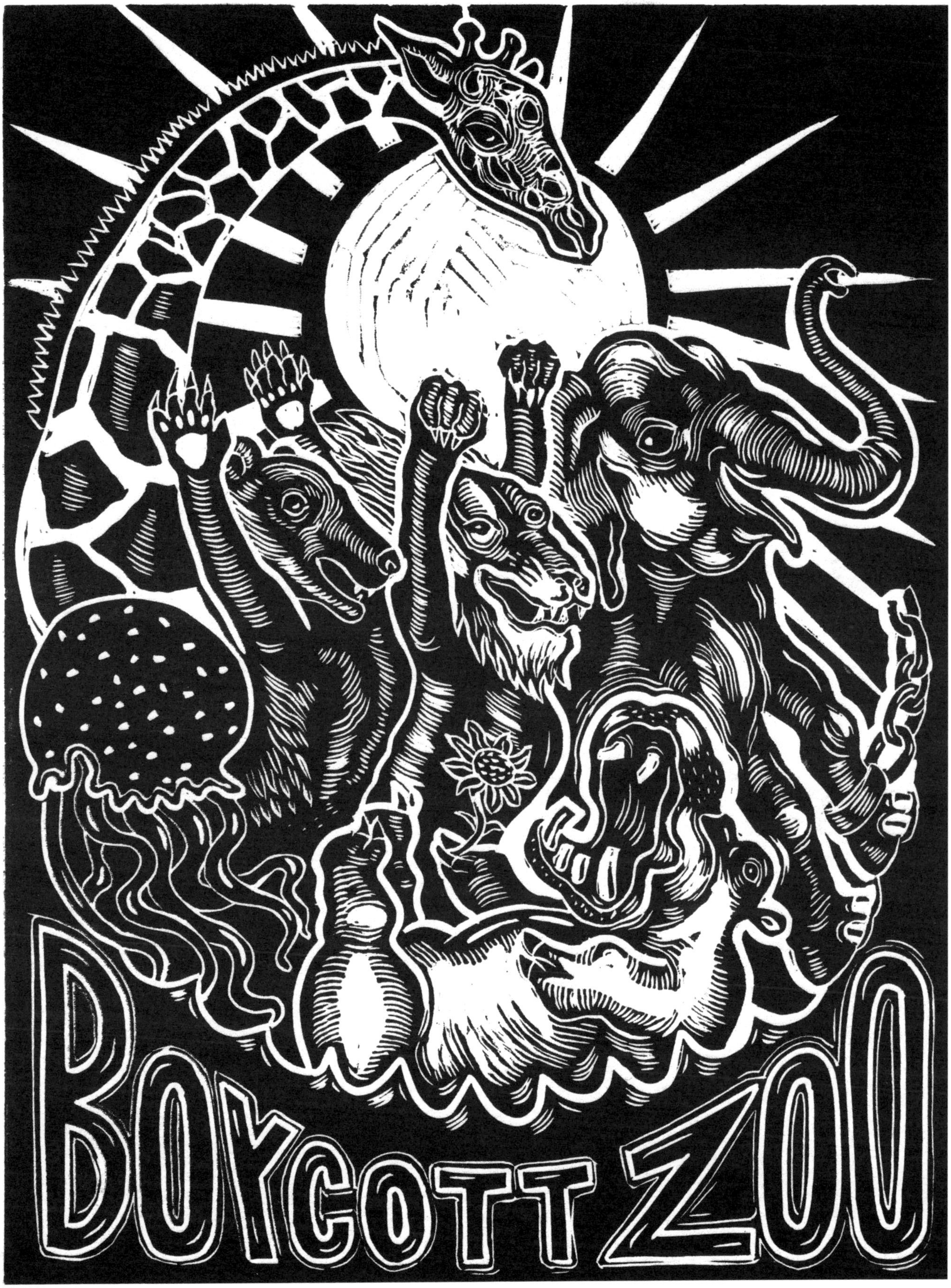
BOYCOTT ZOO

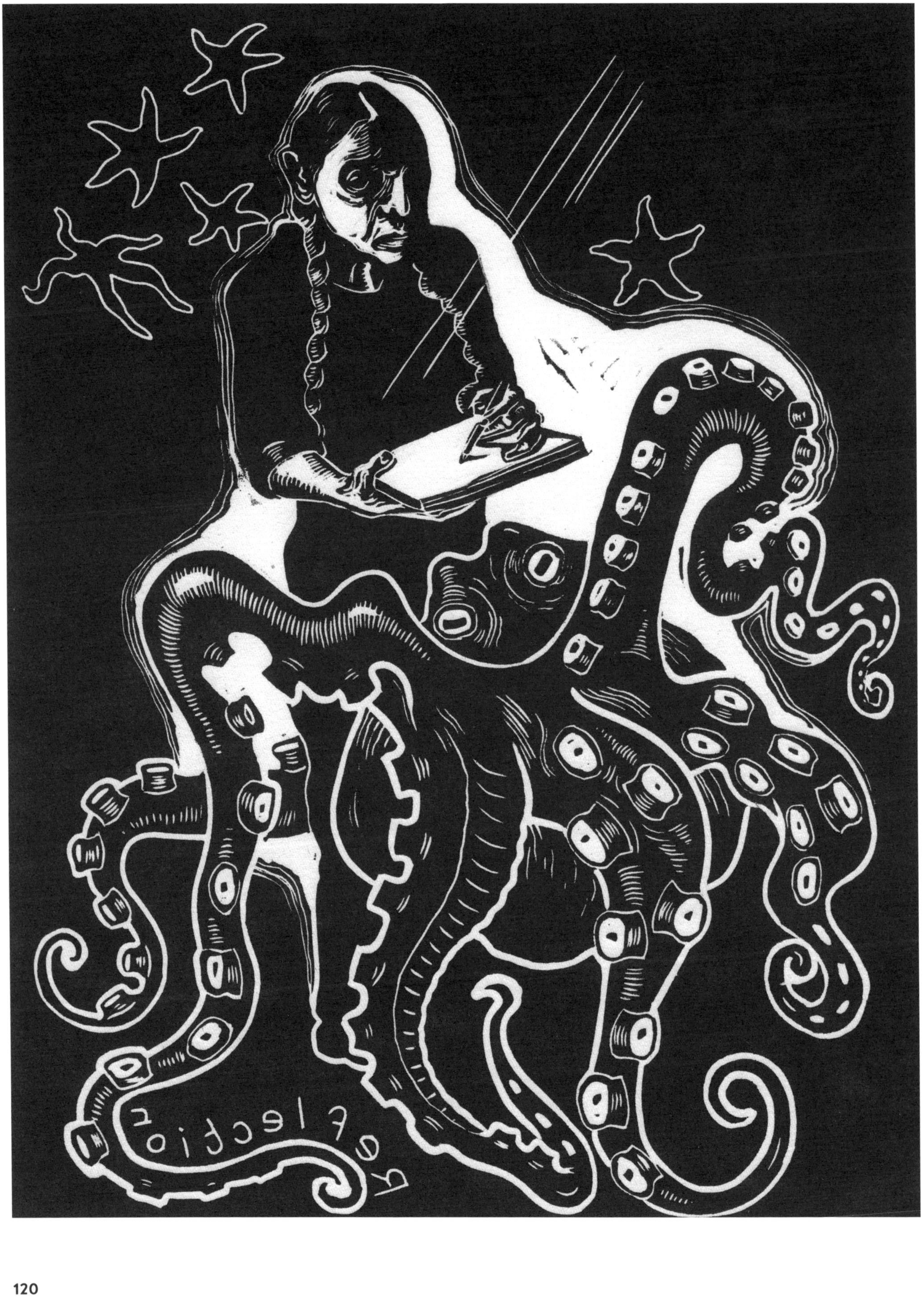

ARTIST'S STATEMENT

Many years past, awash in the dark ages of speciesism, my sister and I visited a large zoo. We were entranced by the chimpanzees. One young chimp was swinging on the bars, but using only one hand and two feet. In the other hand, was gently held a dead mouse. Every so often, the chimp would stop to examine her hidden mouse, looking at the tail and the whiskers, and grooming the silky fur. A prize beyond compare in the barren prison. On the concrete floor, an adult chimp sat catching boiled candy a man was throwing at her. He purchased this bag of candy in the gift store. The chimp would barely move her body, her long arm would shoot out and catch the sweet every time. Each candy was wrapped in a transparent cellophane of different colors, and she would catch the candy, unwrap it, and pop it into her mouth. She put the wrappers under her foot. The man got bored and wandered off to hurl candy at someone else. We stayed and watched, focused on the wrappings underfoot. The chimp lifted her foot and then carefully smoothed out all the wrappings, one at a time, and placed them in a neat pile. She put the first one, which was yellow, up to her eye, and looked into the sky and around the horizon. She then chose a blue wrapper and put it over the yellow, to create a green sky. She did different permutations with all the colors—a purple sky, a red sky, then all of the colors at once. Perhaps, through these colors, she remembered an ancient homeland. Not the grey overcast sky of England, with its horizon of grey concrete, but a jungle canopy alive with birds and plants, and a sky streaked with every color.

Making the drawings for this book, required I revisit zoos. Standing, drawing the prisoners in a sea of cell phones and baby carriages. The living props for human entertainment rarely making eye contact. If the animals don't move, the humans become irritated and demand action by shouting, banging on the glass, or they get bored and move along. What is unusual is people seeing a human, one of their own kind, drawing. One of the rarest animals on earth prompted barely a glance, but drawing that same animal prompted many questions. "Why are you doing this?" "What is it for?" "I wish I could draw." "Look, children, an artist!" One person said sadly, "I am glad you are doing this."

The image of Cleo the octopus in this book was made from sketches. She saw me, and I saw myself, only as a reflection in her tank prison. The octopus confounds science with her intelligence. Her huge brain, in proportion to her body, assures her ability to remember and learn. Memory suggests longevity, but Cleo needs to learn fast—her life span is so short. Barely two years. Cleo can make herself invisible by changing her color to match the environment. She can change her shape to squeeze through the smallest hole. She can recognize the different human faces of her guards. She can use tools. Our only commonality with Cleo's physiognomy is the human tongue, a muscle without skeletal structure. Cleo yearned for freedom; she had stereotypical behaviors. Two minutes of our time could be a hundred years in her time. She doesn't belong trapped in the net of the wicked human gaze. She is not our property, not our entertainment, she is not our "food." I am so sorry, Cleo, this book is for you.

—Sue Coe, June 2018

AK Press is small, in terms of staff and resources, but we also manage to be one of the world's most productive anarchist publishing houses. We publish close to twenty books every year, and distribute thousands of other titles published by like-minded independent presses and projects from around the globe. We're entirely worker-run and democratically managed. We operate without a corporate structure—no boss, no managers, no bullshit.

The Friends of AK program is a way you can directly contribute to the continued existence of AK Press, and ensure that we're able to keep publishing books like this one! Friends pay $25 a month directly into our publishing account ($30 for Canada, $35 for international), and receive a copy of every book AK Press publishes for the duration of their membership! Friends also receive a discount on anything they order from our website or buy at a table: 50% on AK titles, and 20% on everything else. We have a Friends of AK ebook program as well: $15 a month gets you an electronic copy of every book we publish for the duration of your membership. You can even sponsor a very discounted membership for someone in prison.

Email FRIENDSOFAK@AKPRESS.ORG for more info, or visit the Friends of AK Press website: HTTPS://WWW.AKPRESS.ORG/FRIENDS.HTML.

There are always great book projects in the works—so sign up now to become a Friend of AK Press, and let the presses roll!